Christopher D. Rahn

Mechatronic Control of Distributed Noise and Vibration

Springer
Berlin
Heidelberg
New York
Barcelona
Hong Kong
London
Milan
Paris
Singapore
Tokyo

Christopher D. Rahn

Mechatronic Control of Distributed Noise and Vibration

A Lyapunov Approach

With 59 Figures

Springer

Associate Professor CHRISTOPHER D. RAHN
The Pennsylvania State University
Department of Mechanical and Nuclear Engineering
150 A Hammond Building
University Park, PA 16802
USA
e-mail: cdr10@psu.edu

ISBN 3-540-41859-8 Springer-Verlag Berlin Heidelberg NewYork

Cip-data applied for

Springer-Verlag is a company in the BertelsmannSpringer publishing group
http://www.springer.de

Printed in Germany

Typesetting: Camera ready by authors
Cover-Design: design & production, Heidelberg
Printed on acid free paper SPIN: 10834053 62/3020/kk - 5 4 3 2 1 0

To Matthew, Kevin, and Katelin

PREFACE

Vibration and noise reduce the perceived quality, productivity, and efficiency of many electromechanical systems. Vibration can cause defects and limit production speeds during manufacturing and produce premature failure of finished products due to fatigue. Potential contact with a vibrating system or hearing damage from a noisy machine can produce a dangerous, unhealthy, and uncomfortable operating environment.

Recent advances in computer technology have allowed the development of sophisticated electromechanical systems for the control of vibration and noise. The demanding specifications of many modern systems require higher performance than possible with the traditional, purely mechanical approaches of increasing system stiffness or damping. Mechatronic systems that integrate computer software and hardware with electromechanical sensors and actuators to control complex mechanical systems have been demonstrated to provide outstanding vibration and noise reduction. The current trends toward higher speed computation and lower cost, higher performance sensors and actuators indicate the continuing possibilities for this control approach in future applications.

The software control algorithms that act as the brain of the mechatronic system are based on an accurate model of the system to be controlled. At the scale of most mechanical applications, the material composing the system components acts as a continuum. Components that do not deform appreciably under the applied loading may be approximated as rigid bodies. The remaining, distributed components can be modeled by partial differential equations (PDEs) or approximated by ordinary differential equations (ODEs) using numerical discretization techniques such as the Finite Element Method (FEM). Discretization may be the only option for distributed components with complex geometry or built-up assemblies. Many manufacturing, aerospace, acoustic, robotic, transportation, and power transmission applications, however, have the geometric simplicity that make PDE models the most accurate and concise representation of the system dynamics.

The purpose of this book is to present recent advances in the application of Lyapunov's method for distributed parameter systems to the control of vibration and noise. There have been many notable books on vibration and noise control that use a discrete or modal approach (*e.g.* [98, 54, 103, 45, 46, 43, 94]). In graduate engineering textbooks, Inman [54] and Junkins [59] introduce distributed control and Lyapunov theory for distributed systems, respectively. Edited books [125, 23] present contributions to the control of distributed parameter vibration. Much of the mathematical treatises on distributed parameter or infinite dimensional systems (*e.g.* [28]), do not address implementation. The recent book by Luo *et al.* [78] focuses on semigroup theory. In de Queiroz et al. [31], applications of Lyapunov's direct method are presented including distributed parameter systems. Unlike previously published work, this book develops adaptive, output feedback controllers that asymptotically stabilize distributed vibration and noise using few sensors and actuators, compensate for the inertial and electrical actuator dynamics, and learn system parameters such as tension and bending stiffness. Visual feedback control using high speed video, setpoint regulation for systems with rigid body modes, and isolation of bounded disturbances are also presented. The intended audience for the book includes academic and industrial researchers and graduate and advanced undergraduate students in engineering and applied mathematics. In addition to being a reference text, the book is sufficiently broad and self-contained to act as a graduate text in vibration and noise control. It includes modeling, simulation, control development, and implementation for distributed vibration and noise problems.

After a short introduction in Chapter 1, Chapter 2 demonstrates distributed parameter modeling using Hamilton's Principle. The equations of motion of second order (*e.g.* strings, cables, and acoustic ducts) and fourth order (*e.g.* beams and plates) systems are derived. It is shown that the equations of motion for vibration and noise problems can be cast in a matrix operator form with the operators having special properties. Exact solution of the open and closed loop equations of motion using modal analysis is demonstrated. Galerkin's method is used for time simulation of systems with nonuniform parameter distributions and/or damping.

Chapter 3 introduces the mathematical tools needed for the application of Lyapunov's method to distributed parameter systems. Pointwise and L^2 boundedness, asymptotic stability, and exponential stability are defined. Key inequalities are introduced that aid in the bounding of Lyapunov functionals and their time derivatives. Lyapunov's method, Barbalat's lemma, and LaSalle's invariance principle are introduced as stability proof techniques. The basics of semigroup theory are demonstrated to determine the existence and uniqueness of a semigroup solution.

Chapter 4 develops passive controllers for distributed parameter systems. The effects of boundary, distributed, pointwise, and parametric damping are explored. Passive boundary controllers are developed that provide exponential stability despite actuator dynamics. Free boundary problems such as the gantry crane and flexible link robot are studied. Mechatronic implementation experimentally demonstrates the feasibility and performance of the proposed controllers.

Chapter 5 develops Exact Model Knowledge (EMK) controllers for second and fourth order systems. Simple examples of a pinned-controlled string and a clamped-controlled beam are used to demonstrate the techniques. Compensation for complicating effects such as actuator dynamics, nonuniform parameter distributions, and nonlinearities are demonstrated. Complete boundary actuation (*i.e.* on all boundaries) allows setpoint regulation of rigid body and vibrational motion. Isolation controllers for acoustic noise and axially moving beam models are developed. Visual feedback controllers are derived and tested on a flexible link robot using a high-speed video camera and two computers interfaced via a dedicated TCP/IP connection.

Chapter 6 demonstrates how many of the controllers developed in Chapter 5 can be redesigned as adaptive controllers that compensate for parametric uncertainty. Adaptive boundary controllers for nonlinear and spatially varying systems and isolation controllers are developed. Mechatronic testbeds are constructed to demonstrated the improved closed loop performance under EMK and adaptive control relative to passive and classical PID approaches.

I would like to thank my former colleagues and students in the Robotics and Mechatronics Laboratory at Clemson University for their contributions to this book. I was fortunate to have worked with Professor Darren Dawson in the Electrical and Computer Engineering Department without whom many of the approaches presented in this book would never have been developed. My former students, Catalin Baicu, Huseyin Canbolat, Fumin Zhang, Siddhartha Nagarkatti, Yugang Li, Aniket Malatpure, Sushil Singh, and Dan Aron, did much of iterative development of Lyapunov functionals and experimental implementation. I greatly appreciate the review of an early draft of this book by the students in ME 893, *Control of Vibration and Noise* at Clemson in Spring of 2000.

Christopher D. Rahn
University Park, PA

CONTENTS

LIST OF SYMBOLS

$(\cdot,\cdot)$	inner product
$\langle\cdot,\cdot\rangle$	state space inner product
$\|\cdot\|$	norm
$\|\cdot\|_e$	Euclidean norm
$\|\cdot\|_2$	L_2 norm
A	amplitude or area
$\mathcal{A}_0$, $\mathcal{B}_0$, $\mathcal{C}_0$	stiffness, input, and damping operators
$\mathcal{A}$, $\mathcal{B}$	state space operators
$\mathcal{B}$	bulk modulus
b, c	viscous damping coefficient
$C(t)$	Lyapunov functional crossing term
$C^i(D)$	function space with continuous i^{th} derivatives on D
$\mathcal{C}^{n\times m}(D)$	space of complex $n \times m$ matrix functions of $t \in [0,\infty)$ and $x \in D$
δ	variational operator
D	spatial domain
$\mathcal{D}$	domain of an operator
δ_{kl}	$\delta_{kl} = 1$ if $k = l$ and 0 if $k \neq l$.
$E(t)$	system energy
EA	axial stiffness
EI	bending stiffness
$f(t)$	applied force
$\mathbf{f}(t)$	input vector
g	gravitational acceleration
$G(s), H(s)$	transfer functions
$H^k(0,L)$	function space with square integrable k^{th} order derivatives $\forall x \in (0,L)$
$\mathcal{H}$	Hilbert space
$\mathbf{i}$, $\mathbf{j}$, $\mathbf{k}$	inertially fixed cartesian coordinates
$\mathbf{I}$	identity matrix
$\mathcal{I}$	identity operator
j	square root of -1
J	rigid body mass moment of inertia
k	spring stiffness

k_d	damping or derivative control gain
k_τ, k_e	electromechanical torque/force and back-emf constants
L	domain length
$L^2(0, L)$	space of square integrable functions on $(0, L)$
$\mathcal{L}^n_\infty(D)$	space of bounded vector functions $\forall t \in [0, \infty)$ and $x \in D$
λ	decay exponent or eigenvalue
$\lambda_{min}(\mathcal{A}), \lambda_{max}(\mathcal{A})$	minimum and maximum eigenvalues of $\mathcal{A}$
m	rigid body mass
m_i	input vector dimension
M	applied moment
$\mathbf{M}$, $\mathbf{C}$, $\mathbf{K}$, $\mathbf{B}$	mass, damping, stiffness, and input matrices
n	discretization order
$\mathcal{N}$	set of all integers
η_k, $\boldsymbol{\eta}$	kth modal coordinate and modal coordinate vector
$p(x, t)$	pressure
$P(x)$	tension
$\Pi_k(t)$	kinetic energy
$\Pi_p(t)$	potential energy
$\Pi_w(t)$	work energy
$q_k(t)$, $\mathbf{q}(t)$	generalized coordinate and coordinate vector
$\mathbf{r}(\mathbf{x}, t)$	displacement vector
$\mathcal{R}(\mathcal{A})$	range of $\mathcal{A}$
$\mathcal{R}^{n\times m}(D)$	space of real $n \times m$ matrix functions of $t \in [0, \infty)$ and $x \in D$
$\mathcal{R}(\lambda)$	real part of λ
ρ	mass per unit length
s	Laplace variable, variable of integration, or arc length coordinate
$\mathcal{S}^n(D)$	see Definition 8
$\mathcal{S}^n_\mathcal{A}(D)$	see Definition 9
t	time
$T(t)$	semigroup
τ	relaxation time constant
$\tau(t)$	applied torque
$\theta(t)$	rotation angle
$\boldsymbol{\Theta}$, $\hat{\boldsymbol{\Theta}}(t)$, and $\tilde{\boldsymbol{\Theta}}(t)$	parameter actual, estimate, and estimate error vectors
$u(\mathbf{x}, t)$	longitudinal (axial) displacement
v	axial transport speed
$V(t)$	Lyapunov functional
$w(\mathbf{x}, t)$	transverse (lateral) displacement
$\mathbf{w}(\mathbf{x}, t)$	generalized coordinate vector
$\mathbf{W}(t)$	linear regression vector
$\mathbf{W}_k(\mathbf{x})$	k^{th} eigenfunction

ω_k	k^{th} eigenvalue or natural frequency
Ω_n	eigenvalue matrix
$\mathbf{x}$	material position vector or state variable
$\mathbf{z}(t)$	state vector
$\mathcal{Z}$	state Hilbert space

Chapter 1
INTRODUCTION

Many control design tools exist for discretized ordinary differential equation models of vibration and noise systems (*e.g.* observer-based state feedback [83] and H^∞ control [46]). A substantial difficulty in the design of these controllers is the choice of the discretization order. Reduction of the infinite dimensional continuum model to a finite dimensional (N^{th} order) discrete model means that an infinite number of motions are neglected. Typically, modal analysis motivates the model reduction. With sufficient system damping, higher order modes can be neglected if the controller rolls off (*i.e.* the controller gain drops sharply) at high frequency. Choice of N too small results in spillover instability that occurs when the controller, designed for the finite dimensional model, senses and actuates higher order modes, driving them unstable [4]. Reduction of the control gain to eliminate spillover often results in poor performance. Choice of N too large results in a high order compensator that can be difficult and costly to implement.

Control design based on distributed parameter models eliminates control spillover instabilities. The physical displacement, slope, and curvature of the continuum constitute the state variables rather than numerically generated node displacements or modes. Thus, the system model closely links to the underlying mechanics. The resulting controllers are often simple, physically motivated, intuitive, and easier to implement and tune. Unfortunately, the relatively few control techniques for distributed parameter models (*e.g.* Lyapunov techniques [59] and semigroup theory [28]) have not been developed into simple design tools.

Distributed parameter modeling and control approaches favor geometrically simple systems. Unlike discretization, distributed parameter methods require extensive hand derivations that become prohibitively difficult with the increasing number and complexity of the underlying PDEs. Thus, the most relevant applications for this approach consist of a few simple components (*e.g.* rigid bodies, second order components (strings, cables, and rods), and fourth order components (beams and plates)) that are simply connected. On the other hand, commercially available software packages exist that can model physical systems with a high degree of detail. Often, however, these models become so large that important features are obscured by irrelevant detail, clouding the intuition and creativity of the control designer and limiting the development of intuitive, physics-based controllers. Modeling decisions that simplify the geometry of the problem based on the control objectives and the system dynamics enable simple PDE models and low order controllers with robust performance.

Simple distributed parameter models describe many mechatronic systems with applications in manufacturing, space, HVAC, robotics, transportation, and power transmission. Spacecraft, for example, often include long booms that isolate sensitive equipment from main spacecraft bus vibrations caused by thruster firings or control wheel unbalance forces. PDEs based on Euler-Bernoulli beam theory can accurately model the longitudinal and transverse boom vibration coupled with the payload and bus rigid body motion. The model parameters (*e.g.* bending stiffness and mass/length) are nonuniform (*i.e.* vary spatially) if the boom structure changes from base to tip. Lyapunov-based controllers for PDE beam models can require application of distributed forces along the elastic member and therefore an infinite number of actuators [10]. Use of few actuators located on the boundary of the system (*i.e.* boundary control) provides a more practical alternative (See *e.g.* [85, 110, 87, 129, 114, 18, 88]).

During the manufacture of continuous materials (paper, foil, plastic, fabric, wire, and yarn) the vibration, tension, and speed of the moving material is regulated to ensure product quality and maximize productivity. In web handling, wide sheets of thin material move at high speed through a variety of production, finishing, coating, and slitting operations. The web can vibrate out-of-plane (transverse direction) due to aerodynamic and roller eccentricity excitation. Steering the web to maintain centerline tracking involves control of lateral web vibration. Longitudinal tension/speed/vibration control allows aggressive speed trajectories (*e.g.* stop/start motion) without breaking the web. Dancer arms and torqued or pivoted rollers provide control actuation. Similarly, the textile processes of melt spinning, twisting, winding, and texturizing used in the high speed production of lightweight fibers and yarns require tight control of material vibration. For fiber and thin web systems, an axially moving string model consisting of three PDEs coupled to rigid body inertia boundary conditions describe the longitudinal, transverse, and lateral motion. Observer-based state feedback based on a reduced-order, discretized model of the infinite dimensional, axially moving system can result in spillover instabilities [127]. To prevent these instabilities, several researchers pursue a distributed approach including quasi-passive point sensor/actuator control [130], boundary control [71, 117, 44], and Lyapunov approaches [107, 105].

Cables are used in many engineering applications due to their inherent low weight, flexibility, strength, and storability. In marine environments, for example, cables moor buoyant structures, tow vehicles and arrays, tether remotely operated vehicles, and transfer materials ship-to-ship for resupply of naval vessels at sea. Unfortunately, cables can transmit vibration to connected mechanical structures resulting in undesirable or catastrophic effects. The transverse stiffness of a cable depends on its tension and length so a cable structure that spans a long distance can produce large vibrations under relatively small disturbances. This vibration can degrade the performance of the connected electrical and/or mechanical subsystems and ultimately lead to failure. For example, ice covered power lines tend to "gallop" in high winds [34]. The cable suspension Tacoma Narrows bridge exhibited high amplitude oscillations in a steady wind that lead to its collapse [122]. In marine structures, vortices can excite large vibration responses that can significantly degrade performance. In cable hydrophone arrays, vortices shed by the cable may induce a "strumming" effect which can severely degrade the performance of the attached hydrophones. In industrial/construction sites (*e.g.*, nuclear waste-handling facilities, shipyards, airports, automobile plants, new buildings, *etc.*) overhead cranes transport large, heavy payloads. During transit, the payload swings freely making accurate positioning difficult, especially in work sites congested with equipment and personnel. Crane operators often move payloads slowly to minimize the induced swing, reducing overall site efficiency. The PDEs for transverse cable vibration have spatially varying tension and couple to boundary ODEs. Researchers study the stability characteristics of distributed parameter string and wave equation systems with simple boundary feedback [106, 21, 67, 69, 132, 65], geometric nonlinearity [116], and more sophisticated feedback laws [84]. Applications to cable [2, 17, 89] and crane [30, 58] control are also investigated.

Active noise control in Heating, Ventilation, and Air Conditioning (HVAC) ducts allows quiet operation without space consuming insulation. The control system permits the free flow of air with minimal back pressure while canceling the transmission of fan and panel flutter noise. For long ducts, the one-dimensional wave equation with a single, constant parameter PDE coupled with the rigid body and electrical dynamics of the control speaker accurately model the plane wave noise propagation process. Much of the string and wave equation research discussed in the previous paragraph applies to acoustic noise control. Few of the books (*e.g.* [94]) and papers (*e.g.* [7, 26, 25, 108, 109, 52, 8, 68]) on feedforward and feedback control of acoustic noise, however, use Lyapunov theory to derive controllers based on a distributed model.

Robots transport payloads for precision manufacturing and material handling in the electronics, nuclear, and automotive industries. In most cases, accurate positioning of the robot payload is essential. Often, vibration rather than load capacity dictates the size and stiffness of the robot manipulator. Light, flexible links require less power to transport heavy payloads but may vibrate excessively. These vibrations can then feed into the robot position control system and lead to system instability. Although the nonlinear kinematics of large angle motion complicate the system model, robots can be modeled by PDEs for the transverse link vibration coupled to ODEs for the joint inertias and actuator dynamics. Distributed parameter techniques have been applied to the rotating single [86, 75, 76] and multiple [21] flexible link control problem and gantry robots with a flexible link [77, 3]. Many other applications, including magnetic bearing supported rotors for power transmission [92] and fluid power systems with long, flexible transmission lines, can be accurately modeled with a few PDEs and appropriate boundary conditions.

The control objectives for a given vibration or noise problem depend on the application and performance specifications. This book addresses stabilization, setpoint regulation, actuator dynamics compensation, and isolation control problems. Most vibration and noise systems are open loop stable. In the absence of damping, the initial condition response oscillates, neither growing nor decaying with time. Asymptotic stabilization requires controllers that dissipate energy and drive the system response to zero. The best case of global exponential stabilization can be obtained for many linear systems using a control law that requires exact knowledge of the model structure and parameters. Under forcing, undamped systems may resonate, resulting in unbounded response. Natural or feedback damping can eliminate resonances and guarantee bounded response for bounded forcing. With poorly known or slowly time-varying parameters, an adaptive control approach [118] provides stability robustness to parametric uncertainty. The low-pass mechanical and electrical dynamics of typical vibration and noise actuators (*e.g.* electric motors and speakers) filter out the high-frequency components of the control signal. Although high-gain current feedback increases the actuator bandwidth, even "fast" actuator dynamics can interact with sufficiently high frequency link vibration to cause instabilities. Backstepping controllers [63] exactly compensate for the actuator dynamics of hybrid PDE/ODE systems.

Setpoint regulation typically involves flexible systems with rigid body modes. These modes exist when the system has at least one free boundary and the open loop system can translate or rotate as a rigid body without vibration. A setpoint regulation controller damps vibration and translates and/or rotates the system to a desired position and/or angle, respectively. Isolation controllers ensure that part of the system is unaffected by disturbances. When flexible structures are disturbed, vibration control often focuses on isolation of quiet parts of the structure from the disturbances. Confinement of vibrations to relatively unimportant areas in the structure allows high precision pointing and/or positioning in the presence of unknown disturbances. Damping in the uncontrolled material ensures bounded response and bounded control inputs for isolation.

The stability proofs for the controllers discussed in this book are based on Lyapunov theory (Lyapunov's second or direct method) [59] wherein the time derivative of a positive Lyapunov functional is shown to be negative. Inequalities are used extensively in the proofs, leading to sufficient and therefore conservative results. In some cases the Lyapunov functional time derivative is bounded by the scaled negative functional itself, producing an exponential decay. In other cases the time derivative is not sufficiently negative and Barbalat's lemma [118] or LaSalle's invariance principle [78] is invoked. The Lyapunov approach is not limited to linear systems and provides a framework for adaptive and backstepping control solutions that compensate for unknown parameters and actuator dynamics, respectively. Unfortunately, there are no constructive techniques for Lyapunov functional design so considerable ingenuity and perseverance are required. One of the most effective ways of developing Lyapunov functionals is to modify functionals from a closely related problem. This book provides many examples from which to build solutions to new problems.

Mechatronics, or the integration of electromechanical sensors and actuators and real-time software for the control of mechanical systems, plays an important role in the control of distributed vibration and noise. Much of the previous research on mathematical control theory for infinite dimensional systems does not address implementability. The resulting compensators are often infinite dimensional and require infinite numbers of sensors and actuators. Only recently have controllers based on distributed parameter models been experimentally tested to demonstrate functionality, performance, and robustness in practice.

The Lyapunov approach used in this book generally allows the designer to choose actuators but not sensors. Electric motors, speakers, and piezoelectric materials, for example, provide force/torque actuation. The control law resulting from Lyapunov analysis of a distributed system may require measurement of position, slope, curvature, and/or shear and their time derivatives, for example. Position and slope are measured using laser or eddy current displacement probes. Strain gages or load cells measure curvature and shear. Distributed sensing using spatially varying piezoelectric film [15, 126, 70] or high-speed video can provide full state feedback including distributed position, slope, curvature, and shear measurements. In many cases, filtered backwards differenced signals can substitute for velocity measurements.

Computer control provides the convenience of rapid and flexible implementation and precise gain tuning. All of the controllers presented in this book could be implemented as analog circuits but in the experiments we use digital implementation. The controller designs are based on a continuous-time, distributed (infinite bandwidth) model. Digital implementation requires a finite time step and the associated bandwidth limitations. Continuous design with digital implementation is a widely used approach for sophisticated controllers. The effects of discrete time implementation on continuous time predictions of performance and stability can be analyzed [39]. If high frequency instabilities occur in the digital implementation of a distributed model based controller, time discretization rather than control design is the likely culprit.

Chapter 2
DISTRIBUTED PARAMETER MODELS

Model-based control development begins with the derivation of an appropriate model for the distributed vibration or noise system under consideration. In this chapter, we develop hybrid PDE/ODE models for the electromechanical components of mechatronic vibration and noise control systems. The mechanical subsystem involves distributed (PDE) field equations coupled to lumped actuator and sensor equations. The electrical dynamics of the interfacing circuits, actuators, and sensors are typically lumped and described by ODEs.

Use of Hamilton's Principle has two advantages over other methods of generating the equations of motion for mechanical systems. First, the equations derive from the system kinetic, potential, and work energies. The Lyapunov functionals for controls development are also based on the system energy, ensuring consistency between the control and modeling approaches. Second, the approach automatically generates boundary conditions, including the effects of actuator inertia. This chapter applies Hamilton's Principle to second (*e.g.* string and cable) and fourth (*e.g.* beam and plate) order distributed systems including the effects of inertial and electrical actuator dynamics. The equations of motion are cast in a matrix operator form and their properties are studied. Modal analysis solves the exact eigenvalues and eigenfunctions. Galerkin's method is introduced to solve damped eigenvalue problems and allow analysis and design of systems with spatially varying parameters.

2.1 Hamilton's Principle

Hamilton's principle [81] generates the equations of motion of mechanical systems based on the kinetic energy $\Pi_k(t)$, potential energy $\Pi_p(t)$, and work energy $\Pi_w(t)$, as follows

$$\int_{t_1}^{t_2} (\delta\Pi_k - \delta\Pi_p + \delta\Pi_w)\, dt = 0, \tag{2.1}$$

where δ is the variational operator and t_1 and t_2 are two time instants. For vibrating systems, the kinetic energy derives from the translational and rotational velocity of flexible material particles and rigid bodies. Material strain and motion in a conservative force field (*e.g.* gravity) produce potential energy. Control, disturbance, and dissipative forces and torques comprise the work energy.

In general, mechanical systems can be divided into lumped and distributed subsystems. Lumped systems involve rigid bodies with finite degrees of freedom that are mathematically described by time-dependent generalized coordinates and ODEs. The potential and kinetic energies of a lumped parameter component are typically functions of the generalized coordinates and their rates, respectively. Distributed components are infinite dimensional and mathematically represented by PDEs involving functions of space and time. The potential and kinetic energies of a distributed component are functionals involving spatial integrals of the material displacement and velocity, respectively. At sufficiently small length and/or fast time scales, however, all mechanical components are distributed. Application of the rigid body assumption to a specific mechanical system requires engineering judgement based on spectral separation. The motion timescale of the "rigid" body if it is allowed to flex should be orders of magnitude faster than other components in the system. Hence, if the natural frequencies constituting the spectrum of the "rigid" body are orders of magnitude higher than the rest of the system, then the flexing motion can be neglected.

2.2 Lumped Systems

2.2.1 Mechanical Systems

Lumped mechanical systems involve rigid bodies and forces and torques from, for example, springs and dampers. An unconstrained rigid body has six degrees of freedom, three rotations and three translations. In most applications, the rigid body is rotationally and/or translationally constrained along one or more axes. Often, uncoupled and unforced motions along axes orthogonal to the axes of interest are assumed small and neglected. Figure 1 shows the displacement vector $\mathbf{r}(t) = u(t)\mathbf{i} + w(t)\mathbf{j}$, rotation angle $\theta(t)$, and applied force $\mathbf{f}(t)$ and torque $\tau(t)$ for a planar rigid body. The kinetic energy and virtual work for the rigid body are

$$\Pi_k = \frac{1}{2}m\left(\dot{u}^2 + \dot{w}^2\right) + \frac{1}{2}J\dot{\theta}^2, \tag{2.2}$$

$$\delta\Pi_w = \mathbf{f}\delta\mathbf{r} + \tau\delta\theta, \tag{2.3}$$

where dots indicate time differentiation and m and J are the mass and mass moment of inertia of the rigid body, respectively.

The applied force $\mathbf{f}(t)$ and torque $\tau(t)$ could result from control actuation, disturbances, damping, or conservative loading. For control actuation, the applied loading is specified by the control design. If a linear rotary damper is attached to the mass, for example, we obtain $\tau = -b\dot{\theta}$, where b is the rotary damping coefficient. Conservative spring forces or gravity loading can be represented by either work energy or potential energy. For example, the work done by a translational spring in the vertical direction

$$\delta\Pi_w = -kw\delta w = \delta\left(-\frac{1}{2}kw^2\right) = -\delta\Pi_p, \tag{2.4}$$

where k and $\Pi_p = \frac{1}{2}kw^2$ are the spring stiffness and potential energy, respectively.

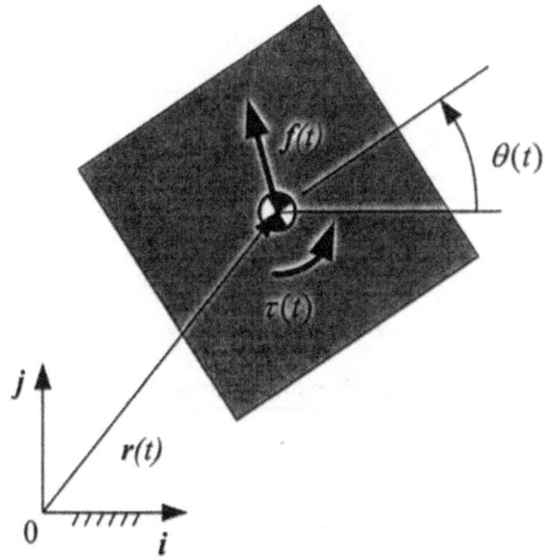

Figure 1 Schematic diagram of a planar rigid body.

2.2.2 Electrical Systems

In many cases, the actuators and sensors in a mechatronic noise or vibration control system are electromechanical devices. DC motor and acoustic loudspeaker actuators, for example, involve rotary and translational motion, respectively. Models of these actuators typically include electrical dynamics associated with the inductance L_a and resistance R of the device as well as the mechanical inertia, stiffness, and damping associated with the rotating or translating output. The input to the electrical dynamics is a voltage $E(t)$ and the output is a torque $\tau(t)$ or force $f(t)$. The electrical dynamics of a loudspeaker are

$$L_a\dot{I} + RI + \kappa_\tau v = E, \tag{2.5}$$

where $I(t)$ is the applied current, κ_τ is the torque constant, and $v(t)$ is the speaker velocity. The electromagnetic force applied to the speaker cone is

$$f = \kappa_\tau I. \tag{2.6}$$

The DC motor model is the same as Eqs. (2.5) and (2.6) with velocity and force replaced with angular velocity and torque.

2.3 Distributed Systems

Second and fourth order distributed mechanical systems involve PDEs with two time derivatives and no more than two and four spatial derivatives, respectively. Examples of second order systems include the transverse and axial vibration of strings and

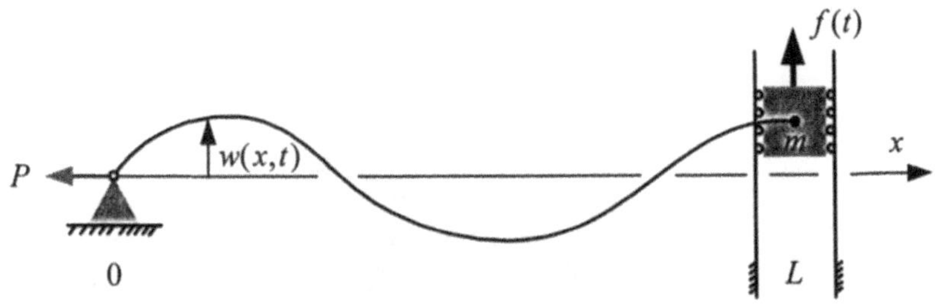

Figure 2 Schematic diagram of a boundary controlled string.

cables, axial and torsional vibration of beams, and noise propagation. Beams and plates are fourth order systems. This section provides example models for second and fourth order distributed mechanical systems that are used later in the book for model-based control development.

2.3.1 Second Order Systems

Boundary Controlled String

Figure 2 shows a schematic diagram of a string vibrating in a transverse plane. The displacement $w(x,t)$ depends on the position variable x and time t. The string has axial tension P, mass per unit length ρ, and length L. A boundary control force is applied to mass m at $x = L$. We assume that the tension is independent of space and time, the mass is constrained to move transversely without rotation, the transverse displacement is small, the out-of-plane and longitudinal vibration are neglected, damping is negligible, and the string is perfectly pinned to the ground at $x = 0$ and to the mass at $x = L$.

The kinetic energy of the system is

$$\Pi_k = \frac{1}{2}mw_t^2(L,t) + \frac{1}{2}\int_0^L \rho w_t^2 dx, \tag{2.7}$$

where the subscript t indicates partial differentiation with respect to time. The first term in Eq. (2.7) comes from the kinetic energy of the mass. The second term represents the kinetic energy of the string. Substitution of Eq. (2.7) in Eq. (2.1) requires evaluation of the integral

$$\begin{aligned} \int_{t_1}^{t_2} \delta\Pi_k dt &= \int_{t_1}^{t_2} \left(mw_t(L,t)\delta w_t(L,t) + \int_0^L \rho w_t \delta w_t dx \right) dt, \\ &= \left[mw_t(L,t)\delta w(L,t) + \int_0^L \rho w_t \delta w dx \right]\Bigg|_{t_1}^{t_2} \\ &\quad - \int_{t_1}^{t_2} \left(mw_{tt}(L,t)\delta w(L,t) + \int_0^L \rho w_{tt} \delta w dx \right) dt, \end{aligned} \tag{2.8}$$

where we interchange the operator δ with $\partial/\partial t$ and integration with respect to t and x and perform integration by parts with respect to t. The variation of w vanishes at $t = t_1$ and $t = t_2$ for all x, so Eq. (2.8) becomes

$$\int_{t_1}^{t_2} \delta\Pi_k dt = -\int_{t_1}^{t_2} \left(m w_{tt}(L,t)\delta w(L,t) + \int_0^L \rho w_{tt} \delta w dx \right) dt. \tag{2.9}$$

The potential energy of the string is

$$\Pi_p = \frac{1}{2}\int_0^L P w_x^2 dx, \tag{2.10}$$

where the subscript x indicates partial differentiation with respect to space. Next we calculate the variation of Eq. (2.10) as follows:

$$\begin{aligned} \int_{t_1}^{t_2} \delta\Pi_p dt &= \int_{t_1}^{t_2}\int_0^L P w_x \delta w_x dx dt \\ &= \int_{t_1}^{t_2} \left(P w_x \delta w|_0^L - \int_0^L P w_{xx} \delta w dx \right) dt, \end{aligned} \tag{2.11}$$

where we interchange the operator δ with $\partial/\partial x$ and integrate by parts with respect to x.

Substitution of Eqs. (2.9) and (2.11) and the virtual work done by the force

$$\delta\Pi_w = f\delta w(L,t) \tag{2.12}$$

into Eq. (2.1) produces

$$\begin{aligned} &\int_{t_1}^{t_2} \left[-\left(m w_{tt}(L,t) + P w_x(L,t) - f \right) \delta w(L,t) + P w_x(0,t)\delta w(0,t) \right. \\ &\left. - \int_0^L (\rho w_{tt} - P w_{xx}) \right] \delta w dx \; dt = 0. \end{aligned} \tag{2.13}$$

The virtual displacements are arbitrary so the integrand of the spatial integral must be zero, producing the field equation

$$\rho w_{tt} - P w_{xx} = 0 \text{ for } x \in (0, L). \tag{2.14}$$

The virtual displacements must satisfy the constraints so $\delta w(0,t) = 0$, eliminating one term from Eq. (2.13). The virtual displacement $\delta w(L,t)$ is not constrained so a natural or force balance boundary condition results at $x = L$. Thus, the boundary conditions for the string are

$$w(0,t) = 0, \tag{2.15}$$

$$m w_{tt}(L,t) + P w_x(L,t) = f. \tag{2.16}$$

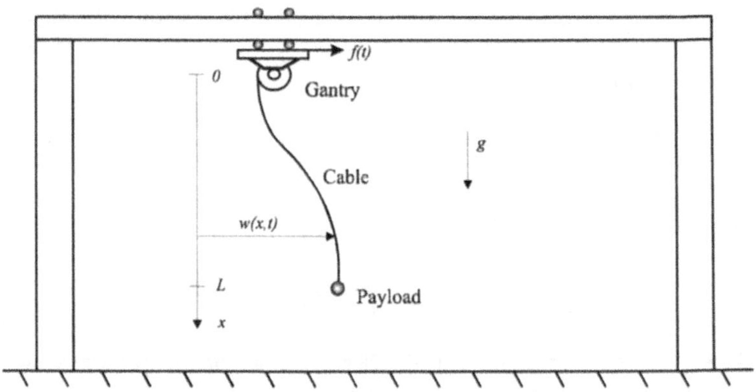

Figure 3 Schematic diagram of a gantry crane model.

Boundary Controlled Gantry Crane

Figure 3 shows a schematic drawing of a planar overhead crane model. The gantry is fixed to a trolley that slides along the top horizontal support. The payload hangs from the gantry on a flexible cable with length L and mass/length ρ. The transverse motion $w(x,t)$ of the cable depends on both the location of the cable particle x and time t. The mass of the gantry and the payload are M_g and M_p, respectively. A control force $f(t)$ is applied to the gantry. We assume that the payload is a point mass, the motion of the gantry along the guide rails is frictionless, the cable slope is small, and the cable is inextensible and perfectly flexible.

From statics and the inextensibility assumption, the tension distributes through the cable as follows:

$$P(x) = g(\rho(L-x) + M_p), \tag{2.17}$$

where g is gravitational acceleration.

The potential energy due to transverse cable displacement is given by:

$$\Pi_p = \frac{1}{2}\int_0^L P\, w_x^2\, dx. \tag{2.18}$$

The kinetic energy includes cable, gantry, and payload terms:

$$\Pi_k = \frac{1}{2}\int_0^L \rho\, w_t^2\, dx \;+\; \frac{1}{2}M_g\, w_t^2(0,t) \;+\; \frac{1}{2}M_p\, w_t^2(L,t). \tag{2.19}$$

The work done by the control force is:

$$\Pi_w = f(t)\, w(0,t). \tag{2.20}$$

Use of Hamilton's principle, Eq. (2.1) results in the field equation

$$\rho w_{tt} - (P w_x)_x = 0 \qquad \text{for } x \in (0, L), \tag{2.21}$$

and the boundary conditions

$$M_g w_{tt} - P w_x = f \qquad \text{at } x = 0, \tag{2.22}$$
$$M_p w_{tt} + P w_x = 0 \qquad \text{at } x = L. \tag{2.23}$$

Domain Controlled Axially Moving String

Figure 4 shows a string or web moving at constant speed v between pinned supports. An actuator consisting of force $f_c(t)$ and mass m controls the string displacement $w(x,t)$ for $x \in (0, x_c)$ where x_c is the actuator location. A disturbance force $f_u(t)$ and boundary displacement $u_L(t)$ act on the uncontrolled subdomain $x \in (x_c, L)$ where L is the total domain length. The kinetic energy of the system is

$$\Pi_k = \frac{1}{2}\rho \int_0^{x_c} (vw_x + w_t)^2 dx + \frac{1}{2}\rho \int_{x_c}^{L} (vw_x + w_t)^2 dx + \frac{1}{2} m w_t^2 (x_c, t), \tag{2.24}$$

where ρ is the mass/length of the string. The kinetic energy of the axial moving string divides into the first two terms in Eq. (2.24) because of the discontinuity in $w_x(x,t)$ at $x = x_c$. The third term in Eq. (2.24) comes from the actuator mass. Under assumptions of small string displacement and perfect flexibility, the system potential energy is

$$\Pi_p = \frac{1}{2} P \int_0^{x_c} w_x^2 dx + \frac{1}{2} P \int_{x_c}^{L} w_x^2 dx, \tag{2.25}$$

where P is the constant string tension. The virtual work done by the external forces is

$$\delta\Pi_w = f_c \delta w(x_c, t) + \int_{x_c}^{L} f_u \delta w dx - \int_0^L c w_t \delta w dx, \tag{2.26}$$

where c is an aerodynamic damping coefficient. Substitution of Eqs. (2.24) – (2.26) into Hamilton's Principle Eq. (2.1) yields the field equation:

$$\rho w_{tt} + 2\rho v w_{xt} + c w_t - P_0 w_{xx} = f_u H(x - x_c), \text{ for } x \in (0, x_c) \cup (x_c, L) \tag{2.27}$$

and the boundary conditions:

$$w(0,t) = 0, \tag{2.28}$$
$$w(L,t) = u_L(t), \tag{2.29}$$
$$w(x_c^-, t) = w(x_c^+, t), \tag{2.30}$$
$$P_0 (w_x^- - w_x^+) + m w_{tt} = f_c, \tag{2.31}$$

where $w_x^- = w_x(x_c^-, t)$, $w_x^+ = w_x(x_c^+, t)$, $H(s)$ is the Heaviside step function, and we assume $P_0 = P - \rho v^2 > 0$.

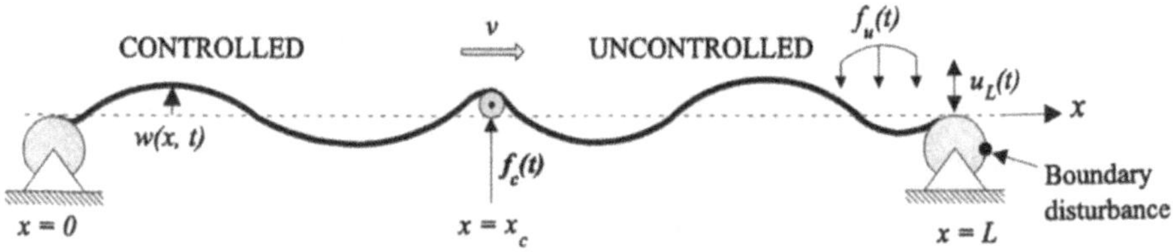

Figure 4 Schematic diagram of an axially moving string system.

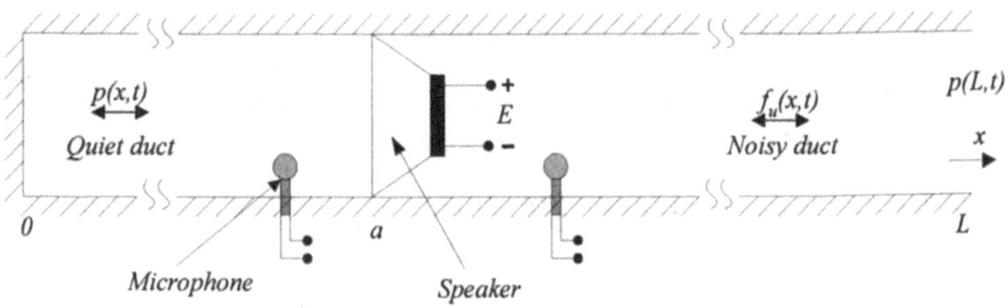

Figure 5 Schematic diagram of a duct noise system.

Domain Controlled Acoustic Duct

Figure 5 shows a finite duct of length L and area A closed at $x = 0$ and with bounded pressure disturbance at $x = L$. The purpose of the loudspeaker, mounted at $x = a$, is to actively isolate the quiet duct $x \in (0, a)$ from the noisy duct $x \in (a, L)$. Microphones measure acoustic pressure on each side of the actuating loudspeaker. The speaker has mass m, damping c, and stiffness k. In practice, the speaker does not completely fill the duct to allow air flow. We assume, however, that the speaker provides an actuated rigid wall between the quiet and noisy ducts. The effects of using a smaller diameter speaker, the flexibility of the speaker cone, and flanking paths through the duct structure are neglected.

The relationship between pressure $p(x,t)$ and displacement $u(x,t)$ in a one-dimensional (*i.e.* planar wave propagation) linearized acoustic system is

$$p(x,t) = \mathcal{B} u_x(x,t), \tag{2.32}$$

where $\mathcal{B}$ is the adiabatic bulk modulus [62]. The kinetic energy of the duct system is

$$\Pi_k = \frac{1}{2}\rho_0 A \int_0^a u_t^2 dx + \frac{1}{2}\rho_0 A \int_a^L u_t^2 dx + \frac{1}{2} m u_t^2 (a,t), \tag{2.33}$$

where ρ_0 is the constant equilibrium density of air. The system potential energy is

$$\Pi_p = \frac{1}{2}\mathcal{B}A\int_0^a u_x^2 dx + \frac{1}{2}\mathcal{B}A\int_0^L u_x^2 dx + \frac{1}{2}ku^2(a,t). \quad (2.34)$$

The virtual work done by the disturbance $f_u(t)$, control force $f(t)$, and viscous dissipation is

$$\begin{aligned}\delta\Pi_w &= f(t)\,\delta u(a,t) - cu_t(a,t)\,\delta u(a,t) + A\int_a^L f_u \delta u dx \\ &\quad - A\int_0^a \tau u_{xt}\delta u_x dx - A\int_a^L \tau u_{xt}\delta u_x dx,\end{aligned}$$

where τ is the relaxation time related to fluid viscosity ([62], pp. 143 - 147).

Using Hamilton's principle (2.1) we obtain the field equation

$$\rho_0 u_{tt} - \tau u_{xxt} - \mathcal{B}u_{xx} = f_u H(x-a), \text{ for } x \in (0,a)\cup(a,L), \quad (2.35)$$

and boundary conditions

$$u(0,t) = 0, \quad (2.36)$$

$$p(L,t) = p_L(t), \quad (2.37)$$

$$u(a^-,t) = u(a^+,t) = u(a,t), \quad (2.38)$$

$$mu_{tt}(a,t) + cu_t(a,t) + ku(a,t) - \frac{A\tau}{\mathcal{B}}(p_t^+ - p_t^-) + A(p^- - p^+) = f(t), \quad (2.39)$$

where $H(x)$ is the Heaviside step function, and $p^- = p(a^-,t)$ and $p^+ = p(a^+,t)$.

Boundary Controlled Axially Vibrating String

The longitudinal web transport model , shown in Figure 6, consists of continuous material of length L, axial stiffness EA, and linear density ρ moving between two controlled rollers. The model applies to strings, beams, membranes, and plates because material bending stiffness and axial cross section do not change the longitudinal dynamics. Control torques are applied to each roller to regulate the material axial displacement $u(x,t)$ which is related by time differentiation to the material speed and by spatial differentiation to the material tension. The derivations of kinetic and work energy are similar to previous examples. Axial strain energy, however, is governed by

$$\Pi_p = \frac{1}{2}\int_0^L EAu_x^2\,dx. \quad (2.40)$$

Use of Hamilton's Principle produces the field equation

$$\rho u_{tt}(x,t) - EAu_{xx}(x,t) = 0, \qquad \forall x \in (0,L), \quad (2.41)$$

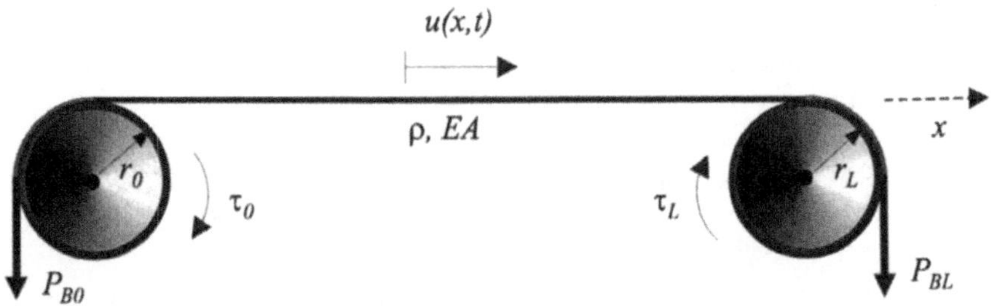

Figure 6 Schematic diagram of an axially moving web.

and boundary conditions

$$mu_{tt}(0,t) - EAu_x(0,t) = f_0(t) \tag{2.42}$$

$$mu_{tt}(L,t) + EAu_x(L,t) = f_L(t), \tag{2.43}$$

where m is the equivalent mass of each roller (assumed equal for simplicity) given by

$$m = \frac{J_0}{r_0^2} = \frac{J_L}{r_L^2}, \tag{2.44}$$

where J_0 and J_L and r_0 and r_L represent the moments of inertia and radii of the two rollers, respectively. The equivalent force control inputs $f_0(t)$ and $f_L(t)$ are related to the control torques $\tau_0(t)$ and $\tau_L(t)$ as follows

$$\begin{aligned} f_0(t) &= \frac{\tau_0(t)}{r_0} - P_{B0}, \\ f_L(t) &= \frac{\tau_L(t)}{r_L} + P_{BL}, \end{aligned} \tag{2.45}$$

where P_{B0} and P_{BL} denote the material tension in the respective adjacent span.

2.3.2 Fourth Order Systems

Boundary Controlled Cantilevered Beam

The boundary controlled Euler-Bernoulli beam shown in Fig. 7 is cantilevered at the left end with a point-mass attached to the right free end. The beam displacement $w(x,t)$ depends on the position x and time t. The beam parameters include the mass/length ρ, bending stiffness EI, and length L. On the controlled boundary, m

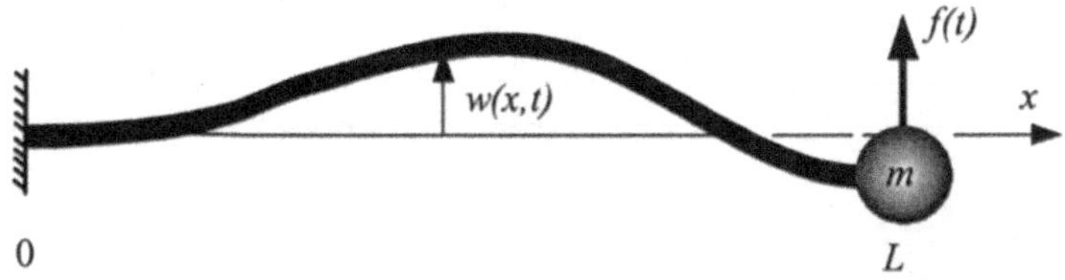

Figure 7 Schematic diagram of a boundary controlled cantilevered beam.

and $f(t)$ denote the attached point mass and boundary control input force, respectively. We assume that the beam is sufficiently thin so that Euler-Bernoulli beam theory applies [81].

The kinetic and work energy of the system are given by Eqs. (2.7) and (2.12), respectively. The potential energy of the beam can include a tension term as in Eq. (2.10) but this is often negligible relative to the bending strain energy given by

$$\Pi_p = \frac{1}{2}\int_0^L EIw_{xx}^2 dx. \tag{2.46}$$

Substitution of the energy terms into Hamilton's Principle (2.1) produces the field equation

$$\rho w_{tt} + EIw_{xxxx} = 0, \tag{2.47}$$

and boundary conditions

$$w(0,t) = w_x(0,t) = w_{xx}(L,t) = 0, \tag{2.48}$$

and

$$mw_{tt}(L,t) = EIw_{xxx}(L,t) + f(t). \tag{2.49}$$

Flexible Link Robot Arm

The flexible link robot arm shown in Fig. 8 is modeled as an Euler-Bernoulli beam clamped to a rotating hub at one end, and attached to an payload mass at the other free end. A control torque input $\tau_h(t)$ is applied to the hub with the dual control objectives of providing a desired angular displacement $w_x(0,t) \to \theta_d$ and regulating the link vibration. Without loss of generality we assume $\theta_d = 0$ so the control objective simplifies to ensuring $w(x,t) \to 0$ as $t \to \infty$.

Based on Euler-Bernoulli beam modeling assumptions and for small displacement, Hamilton's principle can be used to show that the field equation for the beam is given by

$$\rho w_{tt} + EIw_{xxxx} = 0 \quad \forall x \in (0,L), \tag{2.50}$$

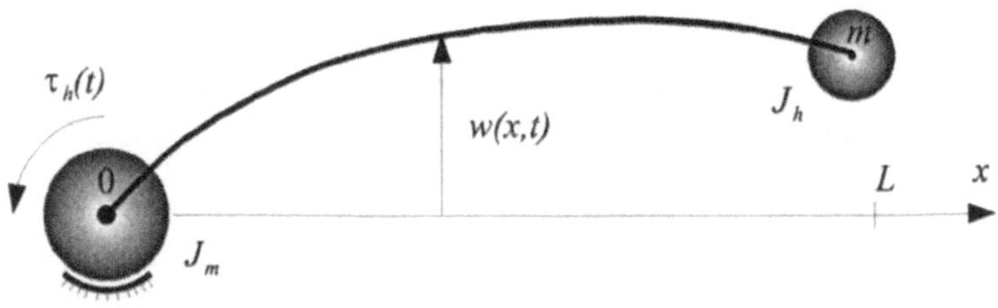

Figure 8 Schematic diagram of a single flexible link robot arm.

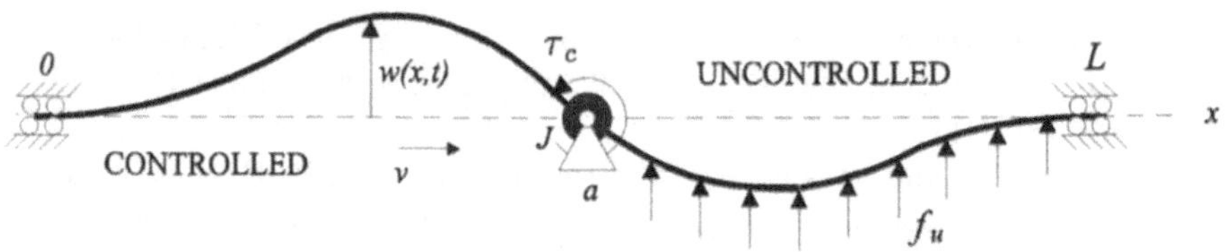

Figure 9 Schematic diagram of an axially moving beam system.

with the following boundary conditions

$$mw_{tt}(L,t) - EIw_{xxx}(L,t) = 0, \tag{2.51}$$

$$J_m w_{xtt}(L,t) + EIw_{xx}(L,t) = 0, \tag{2.52}$$

$$J_h w_{xtt}(0,t) - EIw_{xx}(0,t) = \tau_h(t), \tag{2.53}$$

and

$$w(0,t) = 0, \tag{2.54}$$

where ρ and EI denote the mass/length and bending stiffness of the link, respectively, and m denotes the payload mass. We have also included the rotary inertia of the hub J_h and payload J_m.

Domain Controlled Axially Moving Beam

Figure 9 shows an Euler Bernoulli beam moving at constant speed v between clamped supports. An actuator consisting of torque τ_c and inertia J controls the beam displacement $w(x,t)$ for $x \in [0,a]$ (controlled subdomain), where x is the material position, t is time, and a is the actuator location . A disturbance force f_u acts on the uncontrolled subdomain $x \in (a,L]$ where L is the total span length. Note that despite the pinned condition at $x = a$, the controlled and uncontrolled subdomains are coupled by the beam bending stiffness.

The total system kinetic energy is

$$\Pi_k = \frac{1}{2}\rho\int_0^L (vw_x + w_t)^2\,dx + \frac{1}{2}Jw_{xt}^2(a,t), \tag{2.55}$$

where ρ and v are the constant mass/length and transport speed, respectively. The system potential energy is

$$\Pi_p = \frac{1}{2}\int_0^{a^-} EIw_{xx}^2 dx + \frac{1}{2}\int_{a^+}^{L} EIw_{xx}^2 dx + \frac{1}{2}\int_0^{L} Pw_x^2 dx, \tag{2.56}$$

where EI and P are the constant bending stiffness and tension of the beam, respectively. The virtual work done by the external forces and torque is

$$\delta\Pi_w = \tau_c \delta w_x(a,t) + \int_a^L f_u \delta w dx - \int_0^L c w_t \delta w dx, \tag{2.57}$$

where c is a viscous damping coefficient. We assume distributed relative velocity damping. Other damping models (*e.g.* absolute velocity, material, or boundary damping) could also be assumed.

Substitution of Eqs. (2.55) - (2.57) into Hamilton's Principle (2.1) yields the field equation:

$$\rho w_{tt} + 2\rho v w_{xt} - P_0 w_{xx} + EI w_{xxxx} + c w_t = f_u H(x-a), \ \forall x \in (0,a) \cup (a,L), \tag{2.58}$$

and boundary conditions:

$$\begin{aligned} w(0,t) &= w(L,t) = 0, && (2.59)\\ w_x(0,t) &= w_x(L,t) = 0, && (2.60)\\ w(a,t) &= 0, && (2.61)\\ w_x(a^-,t) &= w_x(a^+,t), && (2.62) \end{aligned}$$

and

$$J w_{xtt} - EI\left(w_{xx}^+ - w_{xx}^-\right) = \tau_c \text{ at } x = a, \tag{2.63}$$

where $w_{xx}^- = w_{xx}(a^-,t)$, $w_{xx}^+ = w_{xx}(a^+,t)$, $P_0 = P - \rho v^2$, and $H(x)$ is the Heaviside step function.

Parametrically Controlled Beam

Figure 10 shows a schematic diagram of a parametrically controlled beam. The beam is clamped at $x = 0$ and $x = L$. The shape of the vibrating beam is described by the longitudinal $u(x,t)$ and transverse $w(x,t)$ displacement fields. At $x = L$, a control actuator applies a longitudinal force $F(t)$ and a sensor measures the longitudinal displacement $y(t) = u(L,t)$.

The kinetic energy of the beam, neglecting rotatory inertia, derives from the longitudinal and transverse velocities

$$\Pi_k = \frac{\rho}{2}\int_0^L u_t^2 + w_t^2 dx, \tag{2.64}$$

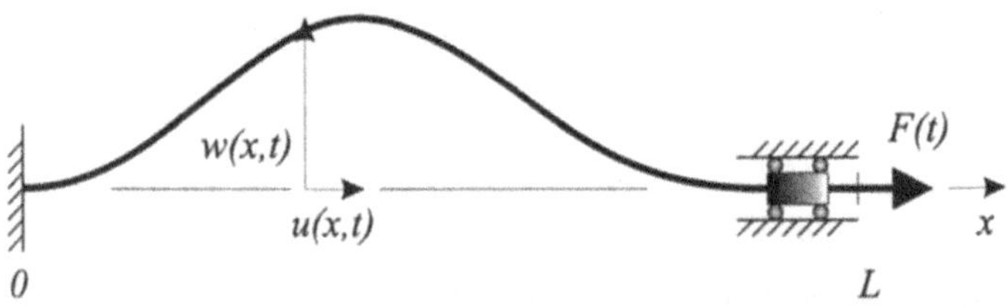

Figure 10 Schematic diagram of a clamped, parametrically controlled beam.

where ρ is the mass/length of the beam. The potential energy, including nonlinear strain deriving from $w(x,t)$ and linear flexure, is

$$\Pi_p = \frac{1}{2}\int_0^L \left[EA(u_x + \frac{1}{2}w_x^2)^2 + EIw_{xx}^2\right] dx, \tag{2.65}$$

where shear deformation has been neglected. The work done by the axial force is

$$W = Fu(L,t). \tag{2.66}$$

Substitution of (2.64) through (2.66) into Hamilton's Principle gives the equations of motion

$$\rho u_{tt} - EA\left(u_x + \frac{1}{2}w_x^2\right)_x = 0, \tag{2.67}$$

$$\rho w_{tt} - EA\left[\left(u_x + \frac{1}{2}w_x^2\right)w_x\right]_x + EIw_{xxxx} = 0, \tag{2.68}$$

and boundary conditions,

$$u(0,t) = 0, \quad EA\left(u_x(L,t) + \frac{1}{2}w_x^2(L,t)\right) = F(t), \tag{2.69}$$

$$w(0,t) = w(L,t) = w_x(0,t) = w_x(L,t) = 0. \tag{2.70}$$

Introduction of the dimensionless variables,

$$x^* = x/L, \quad u^* = u/L, \quad w^* = w/L, \quad t^* = t\sqrt{\rho L^4/EI}, \tag{2.71}$$

$$f^* = fL^3/EI, \quad F^* = FL^2/EI, \tag{2.72}$$

in Eqs. (2.67) - (2.70) and omitting the asterisks gives

$$u_{tt} - v_l^2\left(u_x + \frac{1}{2}w_x^2\right)_x = 0, \tag{2.73}$$

$$w_{tt} - v_l^2\left[\left(u_x + \frac{1}{2}w_x^2\right)w_x\right]_x + w_{xxxx} = 0, \tag{2.74}$$

and

$$u(0,t) = 0, \quad v_l^2\left(u_x(1,t)+\frac{1}{2}w_x^2(1,t)\right) = F(t), \tag{2.75}$$

$$w(0,t) = w(1,t) = w_x(0,t) = w_x(1,t) = 0, \tag{2.76}$$

where the ratio of longitudinal to transverse wave speeds is

$$v_l = \sqrt{AL^2/I}. \tag{2.77}$$

The longitudinal Eqs. (2.73), transverse Eqs. (2.74), and the natural boundary condition (2.75) include nonlinear coupling terms.

If Eqs. (2.73) - (2.76) are linearized about $u = w = 0$,

$$u_{tt} - v_l^2 u_{xx} = 0, \tag{2.78}$$

$$w_{tt} + w_{xxxx} = 0, \tag{2.79}$$

$$u(0,t) = 0, \quad v_l^2 u_x(1,t) = F(t), \tag{2.80}$$

$$w(0,t) = w(1,t) = w_x(0,t) = w_x(1,t) = 0, \tag{2.81}$$

then the axial force $F(t)$ cannot influence transverse vibration $w(x,t)$ because the transverse and longitudinal equations are decoupled.

Using a quasistatic assumption, the complexity of the nonlinear Eqs. (2.73) - (2.76) is avoided but the coupling is retained. The transverse wave speed is assumed to be much smaller than the longitudinal wave speed ($v_l >> 1$) and tension variations are assumed to propagate instantaneously. This assumption is valid because, for most systems, longitudinal wave speeds are at least 100 times faster than transverse wave speeds. Substitution of $u_{tt} = 0$ into Eq. (2.73) gives

$$v_l^2\left(u_x+\frac{1}{2}w_x^2\right)_x = 0, \quad x \in (0,1). \tag{2.82}$$

Integration of (2.82), and use of the boundary conditions (2.75) gives

$$v_l^2\left(u_x+\frac{1}{2}w_x^2\right) = F(t), \quad x \in (0,1). \tag{2.83}$$

Substitution of (2.83) into (2.74) gives the transverse equations of motion

$$w_{tt} - F_0 w_{xx} + w_{xxxx} - F_1(t) w_{xx} = 0, \tag{2.84}$$

where $F(t) = F_0 + F_1(t)$. For large v_l, the measured output is

$$y(t) = u(1,t) = -\frac{1}{2}\int_0^1 w_x^2 dx. \tag{2.85}$$

2.4 Matrix Operator Representation

Linear distributed parameter systems can be cast in matrix operator forms that allow study of their common underlying structure and better understanding of their characteristic responses. For the hyperbolic partial differential equations studied in this book, first or second order matrix operator equations can be developed. Many second order, linear, distributed equations can be written in the following form

$$\ddot{\mathbf{w}} + \mathcal{C}_0\dot{\mathbf{w}} + \mathcal{A}_0\mathbf{w} = \mathcal{B}_0\mathbf{f}, \tag{2.86}$$

where the generalized coordinate vector $\mathbf{w}(\mathbf{x}, t)$ includes the distributed field and dynamic (*i.e.* involving acceleration) boundary variables, $\mathcal{A}_0$, $\mathcal{B}_0$, and $\mathcal{C}_0$ are the stiffness, input, and damping/gyroscopic operators, respectively, and $\mathbf{f}(\mathbf{x}, t) \in \mathcal{R}^{m_i}$ is a vector function of disturbance and control inputs. We assume that $\mathbf{w}(\mathbf{x}, t)$ is an element of a Hilbert space $\mathcal{H}$ with inner product $(\cdot, \cdot)$ and induced norm $\|\mathbf{w}\| = \sqrt{(\mathbf{w}, \mathbf{w})}$. The functions in $\mathcal{H}$ must be sufficiently smooth and satisfy all static boundary conditions.

Definition 1 *Let $\mathcal{H}$ be a vector space. An **inner product** is a mapping $(\cdot, \cdot)$ from $\mathcal{H} \bigoplus \mathcal{H}$ into the set of complex numbers C satisfying*

1. $(\mathbf{w}_1, \mathbf{w}_2) = \overline{(\mathbf{w}_2, \mathbf{w}_1)}$ *for all* $\mathbf{w}_1, \mathbf{w}_2 \in \mathcal{H}$ *(overbar indicates complex conjugate).*
2. $(\alpha\mathbf{w}_1 + \beta\mathbf{w}_2, \mathbf{w}_3) = \alpha(\mathbf{w}_1, \mathbf{w}_3) + \beta(\mathbf{w}_2, \mathbf{w}_3)$ *for all* $\mathbf{w}_1, \mathbf{w}_2 \in \mathcal{H}$ *and* $\alpha, \beta \in C$.
3. $(\mathbf{w}, \mathbf{w}) \geq 0$ *and* $(\mathbf{w}, \mathbf{w}) = 0$ *if and only if* $\mathbf{w} = \mathbf{0}$.

For vibrating systems, the operator $\mathcal{A}_0$ is often symmetric,

$$(\mathbf{w}_1, \mathcal{A}_0\mathbf{w}_2) = (\mathcal{A}_0\mathbf{w}_1, \mathbf{w}_2) \text{ for all } \mathbf{w}_1, \mathbf{w}_2 \in \mathcal{H}, \tag{2.87}$$

and, for systems without rigid body modes, positive definite

$$(\mathbf{w}, \mathcal{A}_0\mathbf{w}) > 0 \text{ for all nonzero } \mathbf{w} \in \mathcal{H}. \tag{2.88}$$

For damped systems, the operator $\mathcal{C}_0$ can be symmetric and positive. Gyroscopic systems, however, often have a skew symmetric $\mathcal{C}_0$.

Equation (2.86) can be rewritten in first order or state form with $\mathbf{z} = [\mathbf{w}, \mathbf{v}]^T \in \mathcal{Z}$ as follows

$$\dot{\mathbf{z}} = \mathcal{A}\mathbf{z} + \mathcal{B}\mathbf{f}, \tag{2.89}$$

where

$$\mathcal{A}\mathbf{z} = \begin{bmatrix} \mathbf{v} \\ -\mathcal{C}_0\mathbf{v} - \mathcal{A}_0\mathbf{w} \end{bmatrix}, \ \mathcal{B}\mathbf{f} = \begin{bmatrix} 0 \\ \mathcal{B}_0\mathbf{f} \end{bmatrix}, \tag{2.90}$$

With the definition of an inner product $\langle \cdot, \cdot \rangle$ that satisfies the conditions in Def. 1, $\mathcal{Z}$ is a Hilbert space. An energy inner product, for example, can be defined as follows

$$\langle \mathbf{z}_1, \mathbf{z}_2 \rangle = \left(\mathcal{A}_0^{1/2} \mathbf{w}_1, \mathcal{A}_0^{1/2} \mathbf{w}_2 \right) + (\dot{\mathbf{w}}_1, \dot{\mathbf{w}}_2), \tag{2.91}$$

where we use the shorthand notation $\mathcal{A}_0^{1/2}$. The first term on the right hand side of Eq. (2.91) typically results from integration by parts of $(\mathbf{w}_1, \mathcal{A}_0 \mathbf{w}_2)$ for symmetric and positive $\mathcal{A}_0$. The domain of $\mathcal{A}_0^{1/2}$, $\mathcal{D}\left(\mathcal{A}_0^{1/2}\right)$, is less restrictive than $\mathcal{D}(\mathcal{A}_0)$, allowing a broader function space for $\mathcal{Z}$.

2.4.1 Second Order Systems

Boundary Controlled String

The boundary controlled string of Eqs. (2.14) - (2.16) can be placed in operator form with $\mathbf{w}(\mathbf{x}, t) = [w(x,t), w(L,t)]^T$ and

$$\mathcal{A}_0 \mathbf{w} = \begin{bmatrix} -\frac{P}{\rho} w_{xx} \\ \frac{P}{m} w_x(L,t) \end{bmatrix}, \; \mathcal{B}_0 \mathbf{f} = \begin{bmatrix} 0 \\ \frac{f}{m} \end{bmatrix}, \; \mathcal{C}_0 \dot{\mathbf{w}} = \mathbf{0}. \tag{2.92}$$

We define $\mathcal{H} = \left\{ [a,b]^T \,|\, a \in H^1(0,L),\, b \in \mathcal{R},\, a(0) = 0,\, b = a(L) \right\}$ where

$$H^k(0,L) = \left\{ f : [0,L] \to R \mid f, f', f'', \cdots, f^{(k)} \in L^2(0,L) \right\}, \tag{2.93}$$

and

$$L^2(0,L) = \left\{ f : [0,L] \to R \mid \int_0^L f^2 dx < \infty \right\}. \tag{2.94}$$

Note that $\mathbf{w}(\mathbf{x}, t) \in \mathcal{H}$ implies that $w(x,t)$ and $w(L,t)$ are bounded for any fixed time t. In the limit as $t \to \infty$, however, they may be unbounded. The inner product

$$(\mathbf{w}_1, \mathbf{w}_2) = \frac{1}{2} \int_0^L \rho w_1 w_2 dx + \frac{1}{2} m w_1(L) w_2(L) \tag{2.95}$$

satisfies the conditions in Def. 1 for positive ρ and m. The operator $\mathcal{A}_0$ can be shown to be symmetric as follows:

$$\begin{aligned} (\mathbf{w}_1, \mathcal{A}_0 \mathbf{w}_2) &= \frac{1}{2} \int_0^L -w_1 P w_{2xx} dx + \frac{1}{2} w_1(L) P w_{2x}(L) \\ &= -\frac{1}{2} P w_1 w_{2x} \Big|_0^L + \frac{1}{2} P w_{1x} w_2 \Big|_0^L \\ &\quad + \frac{1}{2} w_1(L) P w_{2x}(L) + \frac{1}{2} \int_0^L -P w_{1xx} w_2 dx, \end{aligned} \tag{2.96}$$

where integration by parts has been used twice. According to the definition of $\mathcal{H}$, $w_1(0) = w_2(0) = 0$, so the lower evaluations of the first two terms disappear in

Eq. (2.96). Upper evaluation of the first term cancels the third term so only the upper evaluation of the second term and the integral remain. Thus,

$$(\mathbf{w}_1, \mathcal{A}_0\mathbf{w}_2) = \frac{1}{2}\int_0^L -Pw_{1xx}w_2dx + \frac{1}{2}Pw_{1x}(L)w_2(L) = (\mathcal{A}_0\mathbf{w}_1, \mathbf{w}_2)$$

and $\mathcal{A}_0$ is symmetric. Similarly,

$$(\mathbf{w}, \mathcal{A}_0\mathbf{w}) = \frac{P}{2}\int_0^L w_x^2dx \geq 0. \tag{2.97}$$

The pinned boundary condition at $x = 0$ allows use of an inequality derived in Chapter 3:

$$\int_0^L w_x^2dx \geq w^2(x,t) \text{ for all } x \in (0, L).$$

Thus, we obtain the result given in Eq. (2.88) and $\mathcal{A}_0$ is positive definite. We calculate $\mathcal{A}_0^{1/2}$ as follows

$$(\mathbf{w}_1, \mathcal{A}_0\mathbf{w}_2) = \frac{1}{2}\int_0^L \sqrt{P}w_{1x}\sqrt{P}w_{2x}dx = \left(\mathcal{A}_0^{1/2}\mathbf{w}_1, \mathcal{A}_0^{1/2}\mathbf{w}_2\right).$$

Thus, the state space inner product is

$$\begin{aligned}\langle \mathbf{z}_1, \mathbf{z}_2\rangle &= \left(\mathcal{A}_0^{1/2}\mathbf{w}_1, \mathcal{A}_0^{1/2}\mathbf{w}_2\right) + (\mathbf{w}_{1t}, \mathbf{w}_{2t}) \\ &= \frac{1}{2}\int_0^L Pw_{1x}w_{2x} + \rho w_{1t}w_{2t}dx + \frac{1}{2}mw_{1t}(L)w_{2t}(L).\end{aligned} \tag{2.98}$$

If $\mathbf{z}_1 = \mathbf{z}_2 = \mathbf{z}$, we obtain the system energy

$$\langle \mathbf{z}, \mathbf{z}\rangle = \Pi_k + \Pi_p.$$

Gantry Crane

We now develop a matrix operator model for the gantry crane system given by Eq. (2.21) - (2.23). First, we close the loop using the simple feedback law

$$f = -k_p(w(0,t) - q_d) - k_dw_t(0,t) + k_aw_x(0,t). \tag{2.99}$$

This controller has position, velocity, and cable slope feedback terms designed to regulate the gantry position $w(0,t)$ at a desired position q_d and damp vibration in the cable. To simplify the analysis, we introduce the nondimensional parameters

$$x^* = x/L, \quad q_d^* = q_d/L, \quad w^* = w/L, \quad t^* = t\sqrt{\frac{g}{L}}, \quad p_o = 1 + \frac{M_p}{\rho L},$$

$$k_p^* = \frac{Lk_p}{M_g g}, \quad k_d^* = \frac{k_d}{M_g}\sqrt{\frac{L}{g}}, \quad \text{and} \quad k_a^* = \frac{k_a + P(0)}{M_g g},$$

into Eqs. (2.21) through (2.23) with the control law Eq. (2.99) to produce

$$w_{tt} - [(p_o - x)\, w_x]_x = 0 \qquad \text{for } x \in (0,1), \tag{2.100}$$

with boundary conditions

$$w_{tt} + k_d w_t + k_p(w - q_d) - k_a w_x = 0 \qquad \text{at } x = 0, \tag{2.101}$$

$$w_{tt} + w_x = 0 \qquad \text{at } x = 1. \tag{2.102}$$

where the asterisks have been dropped for simplicity.

Eqs. (2.100) - (2.102) can be placed in operator form by defining state vector $\mathbf{w}(\mathbf{x},t) = [w(x,t), w(0,t), w(L,t)]^T$, input $\mathbf{f} = q_d$, and the operators

$$\mathcal{A}_0\mathbf{w} = \begin{bmatrix} -[(p_o - x)\, w_x]_x \\ k_p w(0) - k_a w_x(0) \\ w_x(1) \end{bmatrix}, \mathcal{B}_0\mathbf{f} = \begin{bmatrix} 0 \\ k_p q_d \\ 0 \end{bmatrix}, \mathcal{C}_0\dot{\mathbf{w}} = \begin{bmatrix} 0 \\ k_d w_t(0) \\ 0 \end{bmatrix}. \tag{2.103}$$

We define the function space $\mathcal{H} = \{[a,b,c]^T \mid a \in H^1(0,1),\ b,\ c \in \mathcal{R},\ b = a(0),\ c = a(1)\}$ with an inner product

$$(\mathbf{w}_1, \mathbf{w}_2) = \frac{1}{2}\int_0^1 (p_o - x) w_1 w_2 dx + \alpha w_1(0) w_2(0) + \beta w_1(1) w_2(1). \tag{2.104}$$

Using integration by parts and algebraic simplification, we determine α and β such that $\mathcal{A}_0$ is symmetric as follows

$$\begin{aligned}
(\mathbf{w}_1, \mathcal{A}_0\mathbf{w}_2) &= -\frac{1}{2}\int_0^1 (p_o - x) w_1 \left[(p_o - x)\, w_{2x}\right]_x dx \\
&\quad + \alpha w_1(0)(k_p w_2(0) - k_a w_{2x}(0)) + \beta w_1(1) w_{2x}(1) \\
&= -\frac{1}{2}\int_0^1 [(p_o - x)\, w_{1x}]_x (p_o - x) w_2\, dx + \alpha k_p w_1(0) w_2(0) \\
&\quad - \left(\alpha k_a - \frac{1}{2} p_o^2\right) w_1(0) w_{2x}(0) + \left(\beta - \frac{1}{2}(p_o - 1)^2\right) w_1(1) w_{2x}(1) \\
&\quad + \frac{1}{2}(p_o - 1)^2 w_{1x}(1)\, w_2(1) - \frac{1}{2} p_o^2 w_{1x}(0)\, w_2(0) \\
&= (\mathcal{A}_0\mathbf{w}_1, \mathbf{w}_2) - \left(\alpha k_a - \frac{1}{2} p_o^2\right) w_1(0) w_{2x}(0) \\
&\quad + \left(\beta - \frac{1}{2}(p_o - 1)^2\right) w_1(1) w_{2x}(1) + \left(\frac{1}{2}(p_o - 1)^2 - \beta\right) w_{1x}(1)\, w_2(1) \\
&\quad - \left(\frac{1}{2} p_o^2 - \alpha k_a\right) w_{1x}(0)\, w_2(0) \\
&= (\mathcal{A}_0\mathbf{w}_1, \mathbf{w}_2)
\end{aligned}$$

if $\alpha = p_0^2/(2k_a)$ and $\beta = (p_o - 1)^2/2$. We can also show that $\mathcal{A}_0$ is positive:

$$\begin{aligned}(\mathbf{w}, \mathcal{A}_0\mathbf{w}) &= \frac{1}{2}\int_0^1 (p_o - x)^2 w_x^2 dx - \frac{1}{2}(p_o - x)^2 w_x w\Big|_0^1 \\ &+ \frac{p_0^2}{2k_a} w(0)\left(k_p w(0) - k_a w_x(0)\right) + \frac{(p_o-1)^2}{2} w(1) w_x(1) \\ &= \frac{1}{2}\int_0^1 (p_o - x)^2 w_x^2 dx + \frac{p_0^2 k_p}{2k_a} w^2(0) \geq 0. \end{aligned} \tag{2.105}$$

2.4.2 Fourth Order Systems

Boundary Controlled Beam

The boundary controlled cantilevered beam governed by Eqs. (2.47) - (2.48) can be rewritten using the state vector $\mathbf{w}(\mathbf{x}, t) = [w(x,t), w(L,t)]^T$ in the form of Eq. (2.86) as follows

$$\mathcal{A}_0\mathbf{w} = \begin{bmatrix} \frac{EI}{\rho} w_{xxxx} \\ -\frac{EI}{m} w_{xxx}(L,t) \end{bmatrix}, \ \mathcal{B}_0\mathbf{f} = \begin{bmatrix} 0 \\ f \end{bmatrix}, \ \mathcal{C}_0\dot{\mathbf{w}} = \mathbf{0}. \tag{2.106}$$

If we choose $\mathcal{H} = \{[a, b]^T \,| a \in H^2(0, L),\ b \in \mathcal{R},\ a(0) = 0,\ a_x(0) = 0,\ a_{xx}(L) = 0,$ $b = a(L)\}$ with an inner product

$$(\mathbf{w}_1, \mathbf{w}_2) = \frac{1}{2}\int_0^L \rho w_1 w_2 dx + \frac{1}{2} m w_1(L) w_2(L) \tag{2.107}$$

then the operator $\mathcal{A}_0$ is symmetric

$$\begin{aligned}(\mathbf{w}_1, \mathcal{A}_0\mathbf{w}_2) &= \frac{1}{2}\int_0^L EI w_1 w_{2xxxx} dx - \frac{1}{2} EI w_1(L) w_{2xxx}(L) \\ &= \frac{1}{2}\int_0^L EI w_{1xxxx} w_2 dx - \frac{1}{2} EI w_{1xxx}(L) w_2(L) \\ &= (\mathcal{A}_0\mathbf{w}_1, \mathbf{w}_2) \end{aligned} \tag{2.108}$$

and positive

$$\begin{aligned}(\mathbf{w}, \mathcal{A}_0\mathbf{w}) &= \frac{1}{2}\int_0^L EI w w_{xxxx} dx - \frac{1}{2} EI w(L) w_{xxx}(L) \\ &= \frac{1}{2}\int_0^L EI w_{xx}^2 dx \geq 0. \end{aligned} \tag{2.109}$$

Flexible Link Robot Arm

The flexible link robot arm described by Eqs. (2.50) - (2.54) can be placed in operator form with $\mathbf{w}(\mathbf{x},t) = [w(x,t), w_x(0,t), w(L,t), w_x(L,t)]^T$ and

$$\mathcal{A}_0\mathbf{w} = \begin{bmatrix} \frac{EI}{\rho} w_{xxxx} \\ -\frac{EI}{J_h} w_{xx}(0,t) \\ -\frac{EI}{m} w_{xxx}(L,t) \\ \frac{EI}{J_m} w_{xx}(L,t) \end{bmatrix}, \ \mathcal{B}_0\mathbf{f} = \begin{bmatrix} 0 \\ \frac{\tau_h}{J_h} \\ 0 \\ 0 \end{bmatrix}, \ \mathcal{C}_0\dot{\mathbf{w}} = \mathbf{0}. \tag{2.110}$$

We define $\mathcal{H} = \{[a,b,c,d]^T \,|a \in H^2(0,L),\, b,c,d \in \mathcal{R},\, a(0) = 0,\, b = a_x(0),\, c = a(L),\, d = a_x(L)\}$ with an inner product

$$\begin{aligned}(\mathbf{w}_1, \mathbf{w}_2) &= \frac{1}{2}\int_0^L \rho w_1 w_2 dx + \frac{1}{2} J_h w_{1x}(0) w_{2x}(0) \\ &\quad + \frac{1}{2} m w_1(L) w_2(L) + \frac{1}{2} J_m w_{1x}(L) w_{2x}(L). \end{aligned} \tag{2.111}$$

The operator $\mathcal{A}_0$ is symmetric

$$\begin{aligned}(\mathbf{w}_1, \mathcal{A}_0\mathbf{w}_2) &= \frac{1}{2}\int_0^L EI w_1 w_{2xxxx} dx - \frac{1}{2} EI w_{1x}(0) w_{2xx}(0) \\ &\quad - \frac{1}{2} EI w_1(L) w_{2xxx}(L) + \frac{1}{2} EI w_{1x}(L) w_{2xx}(L) &(2.112)\\ &= \frac{1}{2}\int_0^L EI w_{1xxxx} w_2 dx - \frac{1}{2} EI w_{1xx}(0) w_{2x}(0) \\ &\quad - \frac{1}{2} EI w_{1xxx}(L) w_2(L) + \frac{1}{2} EI w_{1xx}(L) w_{2x}(L) &(2.113)\\ &= (\mathcal{A}_0\mathbf{w}_1, \mathbf{w}_2) &(2.114)\end{aligned}$$

and positive

$$\begin{aligned}(\mathbf{w}, \mathcal{A}_0\mathbf{w}) &= \frac{1}{2}\int_0^L EI w w_{xxxx} dx - \frac{1}{2} EI w_x(0) w_{xx}(0) \\ &\quad - \frac{1}{2} EI w(L) w_{xxx}(L) + \frac{1}{2} EI w_x(L) w_{xx}(L) &(2.115)\\ &= \frac{1}{2}\int_0^L EI w_{xx}^2 dx \geq 0. &(2.116)\end{aligned}$$

Note that due to the presence of rigid body modes, $\mathcal{A}_0$ is not positive definite. We calculate $\mathcal{A}_0^{1/2}$ as follows

$$(\mathbf{w}_1, \mathcal{A}_0\mathbf{w}_2) = \frac{1}{2}\int_0^L \sqrt{EI} w_{1xx} \sqrt{EI} w_{2xx} dx = \left(\mathcal{A}_0^{1/2}\mathbf{w}_1, \mathcal{A}_0^{1/2}\mathbf{w}_2\right),$$

so the state space inner product is

$$\begin{aligned}\langle \mathbf{z}_1, \mathbf{z}_2 \rangle &= \left(\mathcal{A}_0^{1/2}\mathbf{w}_1, \mathcal{A}_0^{1/2}\mathbf{w}_2\right) + (\mathbf{w}_{1t}, \mathbf{w}_{2t}) \\ &= \frac{1}{2}\int_0^L EIw_{1xx}w_{2xx} + \rho w_{1t}w_{2t}dx + \frac{1}{2}J_h w_{1xt}(0)w_{2xt}(0) \\ &\quad + \frac{1}{2}mw_{1t}(L)w_{2t}(L) + \frac{1}{2}J_m w_{1xt}(L)w_{2xt}(L) \end{aligned} \tag{2.117}$$

and $\langle \mathbf{z}, \mathbf{z} \rangle$ equals the system energy.

2.5 Simulation

The eigenvalues dictate many of the important response characteristics of linear vibrating systems. Without damping, the system eigenvalues and eigenfunctions are real. The initial condition response oscillates at the natural frequencies. The forced response can resonate if the input frequency matches a natural frequency, leading to linear growth in the response amplitude with time. The eigenfunctions dictate the spatial shape of the response. In this section, we demonstrate how the eigenvalue problem for undamped problems can be exactly solved for many systems. Eigenvalue analysis becomes significantly more complicated for linear systems with damping or spatially varying parameters. Galerkin's methods is introduced to solve these problems.

2.5.1 Exact Modal Analysis for Undamped Systems

Eigenvalue Problem

An eigenvalue problem is generated from the governing system equations by substituting

$$\mathbf{w}(\mathbf{x}, t) = \mathbf{W}(\mathbf{x}) \exp(j\omega t) \tag{2.118}$$

into Eq. (2.86) with $\mathbf{f} = \mathcal{C}_0\dot{\mathbf{w}} = \mathbf{0}$ to produce

$$\left(\mathcal{A}_0 - \omega^2\mathcal{I}\right)\mathbf{W} = \mathbf{0}, \tag{2.119}$$

where $\mathbf{W}(\mathbf{x})$ is the eigenfunction vector, ω denotes the natural frequency, $\mathcal{I}$ is the identity operator, and $j = \sqrt{-1}$. Solution of the eigenvalue problem in Eq. (2.119) involves determination of the $[\omega_k, \mathbf{W}_k(\mathbf{x})]$ pairs that satisfy Eq. (2.119) and the boundary conditions.

Lemma 2 *The natural frequencies and eigenfunctions of Eq. (2.119) are real if $\mathcal{A}_0$ is symmetric.*

Proof: We assume the pair $[\omega_k, \mathbf{W}_k(\mathbf{x})]$ satisfies Eq. (2.119) and normalize the eigenfunction as follows

$$(\mathbf{W}_k, \mathbf{W}_k) = 1.$$

Use of the inner product produces

$$\omega_k^2 = \frac{(\mathbf{W}_k, \mathcal{A}_0 \mathbf{W}_k)}{(\mathbf{W}_k, \mathbf{W}_k)} = (\mathbf{W}_k, \mathcal{A}_0 \mathbf{W}_k)\,.$$

The complex conjugate property of inner products and the symmetry of $\mathcal{A}_0$ yield

$$\overline{\omega_k^2} = \overline{(\mathbf{W}_k, \mathcal{A}_0 \mathbf{W}_k)} = (\mathcal{A}_0 \mathbf{W}_k, \mathbf{W}_k) = (\mathbf{W}_k, \mathcal{A}_0 \mathbf{W}_k)\,,$$

so $\omega_k^2 = \overline{\omega_k^2}$ and $\omega_k^2 \in \mathcal{R}$. From Eq. (2.119), $\mathbf{W}_k(\mathbf{x})$ is a real vector function.
□

Lemma 3 *If $\mathcal{A}_0$ is symmetric with distinct eigenvalues, then the eigenfunctions of Eq. (2.119) are orthogonal as follows:*

$$(\mathbf{W}_k, \mathbf{W}_l) = \delta_{kl} \quad \forall k,\, l, \tag{2.120}$$
$$(\mathcal{A}_0 \mathbf{W}_k, \mathbf{W}_l) = \omega_k^2 \delta_{kl} \quad \forall k,\, l, \tag{2.121}$$

where $\delta_{kl} = 1$ if $k = l$ and 0 otherwise.

Proof: Assume we have two solution pairs $[\omega_k, \mathbf{W}_k(\mathbf{x})]$ and $[\omega_l, \mathbf{W}_l(\mathbf{x})]$ that satisfy Eq. (2.119). Thus,

$$(\mathcal{A}_0 \mathbf{W}_k, \mathbf{W}_l) = (\mathbf{W}_k, \mathcal{A}_0 \mathbf{W}_l) = \omega_l^2 (\mathbf{W}_k, \mathbf{W}_l)\,, \tag{2.122}$$

using the symmetry of $\mathcal{A}_0$ and

$$(\mathcal{A}_0 \mathbf{W}_k, \mathbf{W}_l) = \omega_k^2 (\mathbf{W}_k, \mathbf{W}_l)\,. \tag{2.123}$$

Subtraction of Eq. (2.123) from Eq. (2.122) yields

$$\left[\omega_l^2 - \omega_k^2\right] (\mathbf{W}_k, \mathbf{W}_l) = 0. \tag{2.124}$$

For distinct eigenvalues, $\omega_l^2 - \omega_k^2 \neq 0$, so $(\mathbf{W}_k, \mathbf{W}_l) = 0$ for $k \neq l$. If we normalize the eigenfunctions such that $(\mathbf{W}_k, \mathbf{W}_k) = 1$, then we obtain the condition in Eq. (2.120). Eq. (2.121) results from $(\mathcal{A}_0 \mathbf{W}_k, \mathbf{W}_l) = \omega_k^2 (\mathbf{W}_k, \mathbf{W}_l)$.
□

Modal Simulation

A finite order simulation model can be developed using the modal expansion

$$\mathbf{w}(\mathbf{x}, t) = \sum_{k=1}^{n} \mathbf{W}_k(\mathbf{x}) \eta_k(t)\,, \tag{2.125}$$

where $\mathbf{W}_k(\mathbf{x})$ and $\eta_k(t)$ are k^{th} orthonormal eigenfunction and modal coordinate, respectively, and n is the discretization order. To obtain the modal equations of motion, we evaluate the inner product

$$(\mathbf{W}_k, \mathbf{w}_{tt} + \mathcal{A}_0\mathbf{w} = \mathcal{B}_0\mathbf{f}) \quad \forall k = 1, .., n. \tag{2.126}$$

Substitution of Eq. (2.125) into (2.126) and use of the orthogonality relations in Eqs. (2.120) and (2.121) produces

$$\ddot{\eta}_k(t) + \omega_k^2\eta_k(t) = f_k(t) \qquad \forall k = 1, 2, ..., n, \tag{2.127}$$

where $f_k(t) = (\mathbf{W}_k, \mathcal{B}_0\mathbf{f}(t))$. These modal equations of motion are written in matrix form as

$$\ddot{\boldsymbol{\eta}} + \boldsymbol{\Omega}_n^2\boldsymbol{\eta} = \mathbf{f}_n, \tag{2.128}$$

where $\boldsymbol{\eta} = [\eta_1, ..., \eta_n]^T$, $\mathbf{f}_n = [f_1, ..., f_n]^T$, and $\boldsymbol{\Omega}_n^2 = \text{diag}\{\omega_k^2\}$. The initial conditions for the simulation are determined by

$$\eta_k(0) = (\mathbf{W}_k, \mathbf{w}(\mathbf{x}, 0)), \tag{2.129}$$

$$\dot{\eta}_k(0) = (\mathbf{W}_k, \mathbf{w}_t(\mathbf{x}, 0)). \tag{2.130}$$

Second Order Systems

Boundary Controlled String For one-dimensional, second order systems with uniform parameters like the boundary controlled string governed by Eqs. (2.14) - (2.16), the eigenfunctions have the form

$$W(x) = A\cos(\beta x) + B\sin(\beta x), \tag{2.131}$$

where A, B, and β are real numbers. The Hilbert Space requires $W(0) = 0$ so $A = 0$. Substitution of Eq. (2.131) into the eigenvalue problem Eq. (2.119) with $\mathcal{A}_0$ defined in Eq. (2.92) yields

$$\left(\mathcal{A}_0 - \omega^2\mathcal{I}\right)\mathbf{W} = \begin{bmatrix} -\frac{P}{\rho}W_{xx}(x) \\ \frac{P}{m}W_x(L) \end{bmatrix} - \omega^2\begin{bmatrix} W(x) \\ W(L) \end{bmatrix} = \begin{bmatrix} 0 \\ 0 \end{bmatrix}. \tag{2.132}$$

The first row of Eq. (2.132) produces

$$\omega = \beta\sqrt{\frac{P}{\rho}}. \tag{2.133}$$

The second row of Eq. (2.132) yields the frequency equation

$$\rho\cos\beta L - m\beta\sin\beta L = 0. \tag{2.134}$$

Eq. (2.134) can be rewritten

$$\cos\beta^* - \alpha\beta^*\sin\beta^* = 0, \tag{2.135}$$

where $\alpha = \frac{m}{\rho L}$ is the ratio of the boundary mass to the string mass and $\beta^* = \beta L$. As $\alpha \to \infty$, we obtain pinned-pinned eigenvalues $\omega_k = k\pi\sqrt{\frac{P}{\rho L^2}}$. As $\alpha \to 0$, we have very small boundary mass relative to string mass, resulting in the pinned-free eigenvalues $\omega_k = k\pi\sqrt{\frac{P}{4\rho L^2}}$. Other cases require numerical solution of Eq. (2.134). Normalization of the eigenfunctions means

$$\begin{aligned}(\mathbf{W}_k, \mathbf{W}_k) &= \frac{\rho}{2}\int_0^L W_k^2 dx + \frac{m}{2}W_k^2(L) = B_k^2\left[\frac{\rho}{2}\int_0^L \sin^2(\beta_k x)dx + \frac{m}{2}\sin^2(\beta_k L)\right] \\ &= \frac{B_k^2}{4}\left(\rho L + m\sin^2\beta_k L\right) = 1,\end{aligned} \tag{2.136}$$

where the frequency Eq. (2.134) has been used. Satisfaction of Eq. (2.136) requires

$$B_k = 2\left(\rho L + m\sin^2\beta_k L\right)^{-1/2}.$$

Numerical simulation with arbitrary boundary forcing can be achieved via Eq. (2.128) with

$$f_k(t) = \left(\mathbf{W}_k, \begin{bmatrix} 0 \\ \frac{f}{m} \end{bmatrix}\right) = \frac{1}{2}W_k(L)f(t).$$

For a damping feedback control law, for example,

$$f = -k_d w_t(L,t) = -2k_d\mathbf{B}_n^T\dot{\boldsymbol{\eta}}, \tag{2.137}$$

where $\mathbf{B}_n^T = \frac{1}{2}[W_1(L), W_2(L), ..., W_n(L)]$. The closed loop discretized equations are

$$\ddot{\boldsymbol{\eta}} + 2k_d\mathbf{B}_n\mathbf{B}_n^T\dot{\boldsymbol{\eta}} + \boldsymbol{\Omega}_n^2\boldsymbol{\eta} = \mathbf{0}. \tag{2.138}$$

Gantry Crane The eigenvalue problem for the gantry crane model given by Eqs. (2.100) - (2.102) with $k_d = 0$ (no damping) and $q_d = 0$ (zero input) is

$$\left(\mathcal{A}_0 - \omega^2\mathcal{I}\right)\mathbf{W} = \begin{bmatrix} -[(p_o - x)W_x]_x \\ k_p W - k_a W_x \\ W_x \end{bmatrix} - \omega^2\begin{bmatrix} W(x) \\ W(0) \\ W(L) \end{bmatrix} = \begin{bmatrix} 0 \\ 0 \\ 0 \end{bmatrix}. \tag{2.139}$$

The operator $\mathcal{A}_0$ is nonuniform so the sinusoidal functions given in Eq. (2.131) are not eigenfunctions of $\mathcal{A}_0$. The variable definition

$$s = 2\omega\sqrt{p_o - x} \tag{2.140}$$

transforms the first row of Eq. (2.139) to

$$s\psi + \psi_s + s\psi_{ss} = 0 \qquad \text{for } s \in (s_0, s_1), \tag{2.141}$$

where

$$s_0 = 2\omega\sqrt{p_o}, \quad s_1 = 2\omega\sqrt{p_o - 1}, \tag{2.142}$$

and $\psi(s) = W(x)$. The solutions of Eq. (2.141) are

$$\psi(s) = C_1 J_0(s) + C_2 Y_0(s), \tag{2.143}$$

where J_0 and Y_0 are Bessel functions of the first and second kind, respectively [1].

The second and third rows of Eq. (2.139) transform to

$$s(k_p - \omega^2)\psi + 2k_a\omega^2\psi_s = 0 \qquad \text{at } s = s_0, \tag{2.144}$$

$$s\psi + 2\psi_s = 0 \qquad \text{at } s = s_1. \tag{2.145}$$

The solvability condition for the equations resulting from substitution of Eq. (2.143) into the boundary conditions (2.144) and (2.145) is

$$\begin{aligned} 0 \ = \ & \frac{1}{4}s_0 s_1 J_0(s_0)Y_0(s_1) - \frac{1}{2}s_0 J_0(s_0)Y_1(s_1) - \frac{1}{2}s_1 k_p\sqrt{p_o}J_0(s_0)Y_0(s_1) \\ & + k_p\sqrt{p_o}J_0(s_0)Y_1(s_1) + \frac{1}{2}k_a\omega s_1 J_1(s_0)Y_0(s_1) - k_a\omega J_1(s_0)Y_1(s_1) \\ & - \frac{1}{4}s_0 s_1\omega Y_0(s_0)J_0(s_1) + \frac{1}{2}s_0\omega Y_0(s_0)J_1(s_1) + \frac{1}{2}s_1 k_p\sqrt{p_o}Y_0(s_0)J_0(s_1) \\ & - k_p\sqrt{p_o}Y_0(s_0)J_1(s_1) - \frac{1}{2}s_1 k_a\omega Y_1(s_0)J_0(s_1) + k_a\omega Y_1(s_0)J_1(s_1) \end{aligned} \tag{2.146}$$

where J_1 and Y_1 are the first order Bessel functions of the first and second kind, respectively. Solution of the eigenfrequencies ω_i and eigenfunctions $U_i(x)$ involves determination of the roots of Eq. (2.146) and the normalized coefficients C_1 and C_2.

Substitution of $\rho = 0$ in the equations of motion produces the two degree-of-freedom rigid link model. The nondimensional constant $p_o \to \infty$ for $\rho = 0$. Thus the field equation (2.100) reduces to

$$w_{xx} = 0 \quad \Rightarrow \quad w_x = \theta(t) \quad \Rightarrow \quad w(x,t) = q(t) + \theta(t)x. \tag{2.147}$$

Substitution of Eq. (2.147) into the boundary conditions (2.101) and (2.102) produces the rigid link equations of motion:

$$q_{tt} + k_d q_t + k_p(q - q_d) - k_a\theta \ = \ 0, \tag{2.148}$$

$$\theta_{tt} + q_{tt} + \theta \ = \ 0. \tag{2.149}$$

The eigenvalues λ associated with Eqs. (2.148) and (2.149) satisfy the characteristic equation

$$\lambda^4 + k_d\lambda^3 + (1 + k_p + k_a)\lambda^2 + k_d\lambda + k_p = 0. \tag{2.150}$$

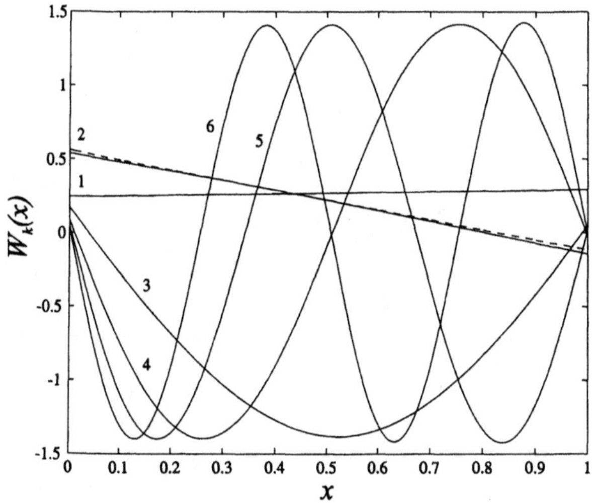

Figure 11 Undamped Flexible (solid) and Rigid (dashed) Gantry Crane Mode Shapes, $p_o = 10$, $k_a = 4$, and $k_p = 1$. Mode numbers are indicated.

Routh's criterion shows that the eigenvalues are asymptotically stable if the control gains are positive. If $k_d = 0$, the eigenvalues are purely imaginary.

We can now optimize the gains for the crane system using a standard algorithm that minimizes the maximum real part of the roots of Eq. (2.150). A minimum is found by the optimizer is near $k_a = 4$, $k_p = 1$, and $k_d = 4$ with a maximum real part of the eigenvalues equal to -1.

Figure 11 shows the first six mode shapes for a relatively rigid cable ($p_o = 10$). The optimal rigid cable proportional and angle gains are used ($k_a = 4$ and $k_p = 1$). Dashed lines indicate the rigid cable mode shapes. The first rigid and flexible mode shapes match closely. The second modes match least well near the boundaries. The third through sixth mode shapes resemble a pinned string. In Figure 12, the reduced cable tension ($p_o = 2$) introduces modal distortion. The first rigid and flexible modes are only slightly changed from the previous case. The rigid second mode deviates significantly from the flexible mode. Higher modes involve substantial motion at the boundaries and increase in amplitude with increasing x. Figure 13 shows the effect of reduced angle gain. Selection of $k_a = 1$ reduces the modal displacements at the gantry ($x = 0$), reducing the coupling between the cable and gantry motion.

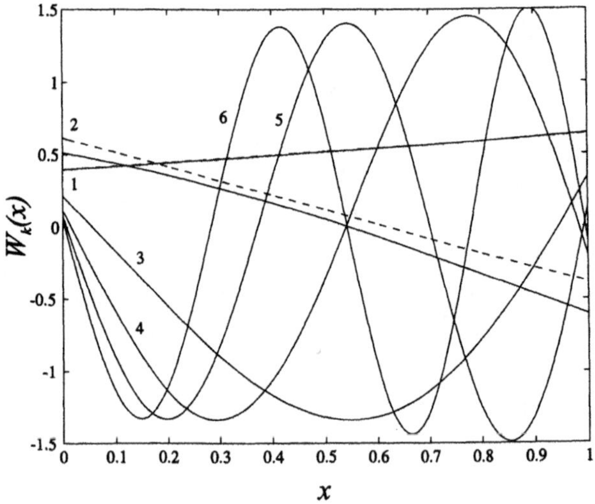

Figure 12 Undamped Flexible (solid) and Rigid (dashed) Gantry Crane Mode Shapes, $p_o = 2$, $k_a = 4$, and $k_p = 1$. Mode numbers are indicated.

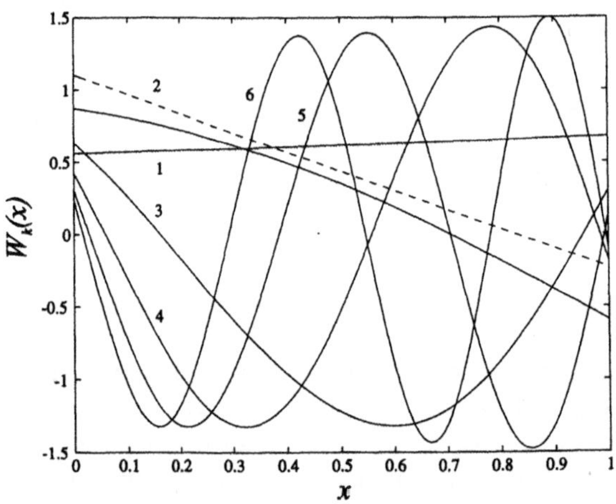

Figure 13 Undamped Flexible (solid) and Rigid (dashed) Gantry Crane Mode Shapes, $p_o = 2$, $k_a = 1$, and $k_p = 1$. Mode numbers are indicated.

Fourth Order Systems

Flexible Link Robot Arm For the Flexible Link Robot Arm, we substitute Eq. (2.118) into Eq. (2.50), to obtain

$$\omega = \sqrt{\frac{EI}{\rho}}\beta^2, \tag{2.151}$$

where the eigenfunctions that satisfy Eqs. (2.50) – (2.54) are of the form

$$W(x) = C_1 \sin(\beta x) + C_2 \cos(\beta x) + C_3 \sinh(\beta x) + C_4 \cosh(\beta x). \tag{2.152}$$

The coefficients C_1, C_2, C_3, and C_4 (to within a scaling constant) and eigenvalues are obtained by substitution of Eq. (2.152) into the boundary conditions (2.54) – (2.52) with $\tau_h = 0$. The resulting characteristic equation

$$\begin{aligned} 0 \;=\; & -mJ_hJ_m\beta^7 - \rho^2 J_h\beta^3 \\ & -2\rho^2 m\beta \sin(\beta L)\sinh(\beta L) \\ & +\left(\rho J_h J_m\beta^6 + \rho m J_m\beta^4 + \rho m J_h\beta^4 - \rho^3\right)\sin(\beta L)\cosh(\beta L) \\ & +\left(\rho J_h J_m\beta^6 - \rho m J_m\beta^4 - \rho m J_h\beta^4 + \rho^3\right)\cos(\beta L)\sinh(\beta L) \\ & +\left(mJ_hJ_m\beta^7 - 2\rho^2 J_m\beta^3 - J_h\rho^2\beta^3\right)\cos(\beta L)\cosh(\beta L) \end{aligned} \tag{2.153}$$

is solved numerically for the eigenvalues β.

2.5.2 Approximate Simulation - Galerkin's Method

For cases involving damping and/or gyroscopic effects or prohibitively complex parameter nonuniformity, an approximate simulation model may be required. Many approximate methods can be used to discretize the governing equations including Finite Difference, Assumed Modes, Finite Element, and Galerkin methods. The Finite Difference method is the least sophisticated approach and can have convergence problems. The finite element method works well for systems with complex geometry. The Assumed Modes method and Galerkin's method use functions that distribute over the entire spatial domain. The Assumed Modes method requires that a variational form of the governing equations be obtained. Galerkin's method works directly on the governing PDEs. In general, both of these methods require significantly fewer degrees of freedom for a given accuracy than the Finite Difference or Finite Element methods. Thus, for the geometrically simple problems studied in this book, Galerkin's method provides accurate, lower order simulations that are easier and faster to compute numerically than other methods.

General Formulation

Galerkin's method assumes a solution in the form of a series of comparison functions. Comparison functions $\hat{\mathbf{W}}_k(\mathbf{x})$ satisfy all of the boundary conditions and are sufficiently differentiable such that $\hat{\mathbf{W}}_k(\mathbf{x}) \in D(\mathcal{A}_0)$. Thus, we expand

$$\mathbf{w}_n(\mathbf{x},t) = \sum_{k=1}^{n} \hat{\mathbf{W}}_k(\mathbf{x}) q_k(t), \tag{2.154}$$

where $q_k(t)$ are the generalized coordinates. Galerkin's method requires that the weighted error over the domain be zero. The weighting functions are chosen to be $\hat{\mathbf{W}}_k(\mathbf{x})$ and the error is calculated using the inner product and Eq. (2.86) as follows

$$\left(\hat{\mathbf{W}}_k, \ddot{\mathbf{w}}_n + \mathcal{C}_0\dot{\mathbf{w}}_n + \mathcal{A}_0\mathbf{w}_n - \mathcal{B}_0\mathbf{f}\right) = 0. \tag{2.155}$$

Definition of

$$\mathbf{M}(k,l) = \left(\hat{\mathbf{W}}_k, \hat{\mathbf{W}}_l\right), \tag{2.156}$$

$$\mathbf{C}(k,l) = \left(\hat{\mathbf{W}}_k, \mathcal{C}_0\hat{\mathbf{W}}_l\right), \tag{2.157}$$

$$\mathbf{K}(k,l) = \left(\hat{\mathbf{W}}_k, \mathcal{A}_0\hat{\mathbf{W}}_l\right), \tag{2.158}$$

$$\mathbf{f}_n(k) = \left(\hat{\mathbf{W}}_k, \mathcal{B}_0\mathbf{f}\right), \tag{2.159}$$

yields the discretized equations

$$\mathbf{M}\ddot{\mathbf{q}} + \mathbf{C}\dot{\mathbf{q}} + \mathbf{K}\mathbf{q} = \mathbf{f}_n. \tag{2.160}$$

Gantry Crane

To generate a discretized simulation model that includes damping for the gantry crane, we introduce the truncated modal series

$$\mathbf{w}_n(\mathbf{x},t) = \sum_{k=1}^{n} \hat{\mathbf{W}}_k(x) q_k(t), \tag{2.161}$$

using n undamped modes $\hat{\mathbf{W}}_k(x)$ that have been normalized according to Eq. (2.120) and generalized coordinates $q_k(t)$. The discretized equations of motion are

$$\ddot{\mathbf{q}} + k_d\mathbf{D}\dot{\mathbf{q}} + \mathbf{\Omega}_n^2\mathbf{q} = \mathbf{G}q_d, \tag{2.162}$$

where the matrix elements

$$\mathbf{D}(k,l) = \frac{p_o}{k_a} W_k(0) W_l(0), \tag{2.163}$$

$$\mathbf{G}(k) = k_p W_k(0). \tag{2.164}$$

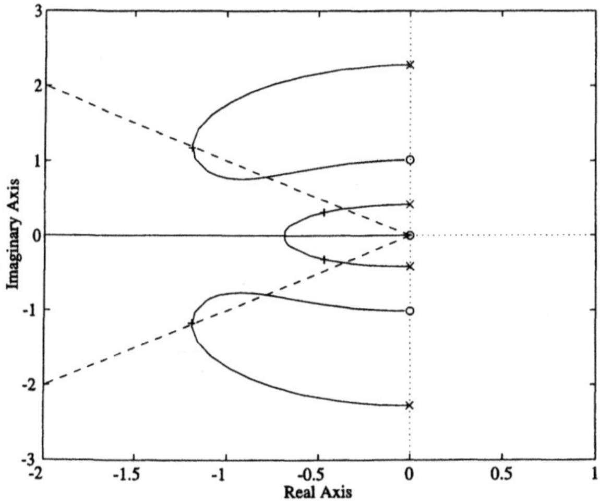

Figure 14 Gantry Crane Root Locus versus k_d: $p_o = 10$, $k_a = 4$, and $k_p = 1$. Crosses indicate the design point ($k_d = 3.8$, $\lambda = -1.2 \pm 1.2i$, $-.47 \pm .31i$).

Note that $\Omega_n = \Omega_n^T > 0$ and $\mathbf{D} = \mathbf{D}^T > 0$, ensuring asymptotic stability of the discrete approximation. The closed loop eigenvalues of the system satisfy

$$\left|s^2\mathbf{I} + k_d\mathbf{D}s + \mathbf{\Omega}_n^2\right| = 0. \tag{2.165}$$

Figure 14 shows the root locus of the closed loop system versus the damping control gain k_d for the modes shown in Fig. 11. The first four poles are asymptotically stabilized by the controller. The dotted line on the root locus plot corresponds to damping of $1/\sqrt{2}$. A good design of $k_d = 3.8$ places the higher frequency pole on the damping line. The lower frequency pole attains a maximum damping of 85% at this gain. Higher frequency poles (not shown) are also stabilized by the controller but with little damping. The root locus matches the rigid cable root locus (not shown) fairly accurately. The rigid cable loci coalesce at -1 with $k_d = 4$.

Figure 15 shows the root locus for the modes shown in Fig. 12. This case corresponds to a lightly loaded cable and high angle gain. Although the root locus looks similar to Fig. 14 for the first four poles, the higher frequency poles experience significantly more damping. The design point shown ensures critical damping on the first mode, 46% damping on the second mode, and 7% damping on the third mode. Figure 16 shows the root locus corresponding to the modes in Fig. 13. The reduced angle gain reduces the damping on the third mode to 2%. A design point on the damping line results in 26% damping on the second mode.

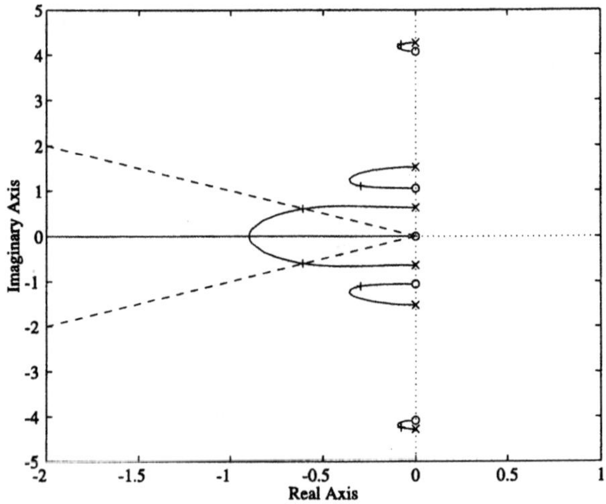

Figure 15 Root Locus versus k_d: $p_o = 2$, $k_a = 4$, and $k_p = 1$. Crosses indicate the design point ($k_d = 4.2$, $\lambda =$ -.57, -.67 $\pm$ 1.3i, -.31 $\pm$ 4.5i).

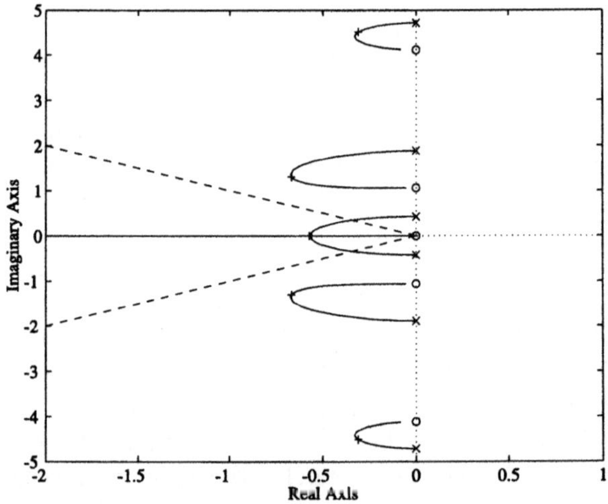

Figure 16 Gantry Crane Root Locus versus k_d: $p_o = 2$, $k_a = 1$, and $k_p = 1$. Crosses indicate the design point ($k_d = 2.1$, $\lambda =$ -.6 $\pm$ .6i, -.3 $\pm$ 1.1i, -.07 $\pm$ 4.2i).

Boundary Controlled Cable

If the boundary controlled string governed by Eqs. (2.21) - (2.23) is oriented perpendicular to the gravity vector and permitted to sag, the tension varies along the length according to

$$P(S) = \sqrt{P_0^2 + \left[\rho g \left(S - \frac{L}{2}\right)\right]^2}, \tag{2.166}$$

where S is the arc length coordinate, g is gravitational acceleration, and P_0 is the tension at the mid-span, $s = L/2$ [101]. The complexity of the spatially varying tension function makes exact modal solutions unobtainable. Introduction of the nondimensional (lower case) variables

$$s = \frac{S}{L}; \quad t = T\sqrt{\frac{g}{L}}; \quad w = \frac{W}{L}; \quad m = \frac{M}{\rho L}; \quad f = \frac{F}{\rho g L};$$
$$p(s) = \frac{P(S)}{\rho g L} = \sqrt{p_0^2 + \left(s - \frac{1}{2}\right)^2}; \quad p_0 = \frac{P_0}{\rho g L} \tag{2.167}$$

into the boundary controlled string field Eqs. (2.14) - (2.16) produces

$$w_{tt} - [p(s)w_s]_s = 0 \tag{2.168}$$

with boundary conditions

$$w(0) = 0, \tag{2.169}$$
$$m w_{tt}(1) + p(1)w_s(1) = f(t). \tag{2.170}$$

If we use a feedback law similar to the gantry crane example, the following closed loop operators for $\mathbf{w}(s,t) = [w(s,t), w(1,t)]^T$ are obtained

$$\mathcal{C}_0\dot{\mathbf{w}} = \begin{bmatrix} 0 \\ k_d/m \end{bmatrix}, \tag{2.171}$$
$$\mathcal{A}_0\mathbf{w} = \begin{bmatrix} -[p(s)w_s]_s \\ (k_p w(1) + (p(1) + k_c)\, w_s(1))\,/m \end{bmatrix}, \tag{2.172}$$

where k_p, k_d, and k_c are nondimensional control gains. For comparison functions we use the eigenfunctions for the boundary controlled string

$$\hat{\mathbf{W}}_k = \begin{bmatrix} B_k \sin(\beta_k s) \\ B_k \sin(\beta_k) \end{bmatrix}, \tag{2.173}$$

where the B_k are normalized according to Eq. (2.136) and the β_k satisfy the frequency Eq. (2.134). From Eqs. (2.156) - (2.158) using the nondimensionalized inner product

Eq. (2.95) we obtain

$$\mathbf{M}(k,l) = \delta_{kl}, \tag{2.174}$$

$$\mathbf{C}(k,l) = k_d \hat{W}_k(1)\hat{W}_l(1), \tag{2.175}$$

$$\mathbf{K}(k,l) = -\frac{1}{2}\int_0^1 \hat{W}_k[p(s)\hat{W}_{l_s}]_s ds \tag{2.176}$$

$$+\frac{1}{2}\hat{W}_k(1)\left[k_p\hat{W}_l(1) + (p(1)+k_c)\,\hat{W}_{l_s}(1)\right], \tag{2.177}$$

where the elements of the stiffness matrix in Eq. (2.176) are numerically generated.

The effectiveness of the cable boundary controller is demonstrated by numerical simulations for a cable system defined by $p_0 = 0.01$ and $m = 0.015$. The response of the first six modes to an unit impulse input applied at the cable mid-span ($s = \frac{1}{2}$) is simulated. The dotted line in Figure 17 shows the performance of the passive boundary controller ($k_p = k_c = 0$). The nondimensional cable displacement at the end-point ($s = 1$) and mid-point ($s = 1/2$) are shown in frames (a) and (b), respectively. The response decays asymptotically to zero with a 15 s/s settling time. The damping is least effective on the higher modes.

The improved damping provided by the full controller ($k_p = k_c = 20$) is shown as the solid lines in Fig. 17(a) for the cable end-point and in Fig. 17(b) for the cable mid-point motion. The angle feedback allows more rapid vibration damping, including higher modes. The response decays asymptotically to zero in 7 s/s, twice as fast as the passive controller. Figure 17(c) shows the nondimensional control force for the two boundary controllers.

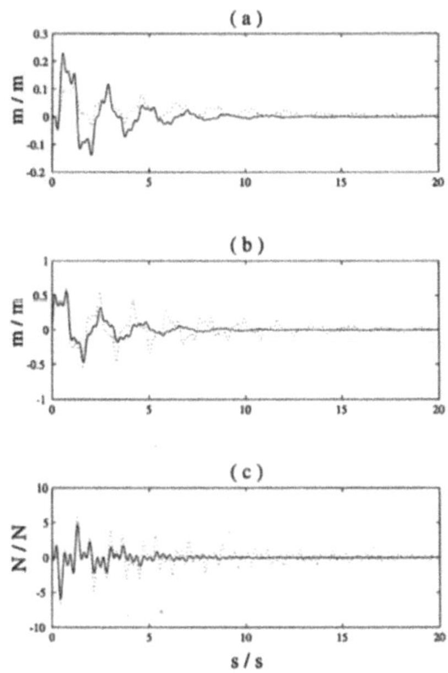

Figure 17 Cable Boundary Controller Impulse Response Simulation: $\cdots$ Damping ($k_d = 10$, $k_p = 0$, $k_c = 0$); – Full Control ($k_d = 10$, $k_p = 20$, $k_c = 20$) (a) Mid-Point Displacement, $w(1/2, t)$; (b) End-Point Displacement, $w(1, t)$; (c) Control Effort, $f(t)$.

Chapter 3
MATHEMATICAL PRELIMINARIES

This chapter introduces stability definitions, inequalities, semigroup theory, and stability theorems for distributed systems. These mathematical tools are used extensively in the following chapters to develop passive, exact model knowledge, adaptive, and backstepping controllers for distributed vibration and noise.

3.1 Definitions

We give the following definitions of boundedness or stability, asymptotic stability, and exponential stability. These definitions are termed "strong" because they require pointwise rather than normed stability so all displacements in the domain D must satisfy the specified condition. They are also global stability definitions because neither the size of the displacement nor the initial conditions $\mathbf{z}(\mathbf{x},0) \in \mathcal{Z}$ are limited.

Definition 4 *All n elements of the multivariable vector function $\mathbf{f}(\mathbf{x},t) \in \mathcal{L}_\infty^n(D)$ are **bounded** $\forall \mathbf{x} \in D$ and $\forall t \in [0,\infty)$.*

Definition 5 *The position $\mathbf{w}(\mathbf{x},t) = 0$ is defined to be **strongly stable** if $\mathbf{w}(\mathbf{x},t) \in \mathcal{L}_\infty(D)$ for any $\mathbf{z}(\mathbf{x},0) \in \mathcal{Z}$.*

Bounded scalar functions of time $f(t)$, n-element vector functions of time $\mathbf{f}(t)$, scalar functions of space and time $f(\mathbf{x},t)$, and n-element vector functions of space and time $\mathbf{f}(\mathbf{x},t)$ are indicated by $f(t) \in \mathcal{L}_\infty$, $\mathbf{f}(t) \in \mathcal{L}_\infty^n$, $f(\mathbf{x},t) \in \mathcal{L}_\infty(D)$, and $\mathbf{f}(\mathbf{x},t) \in \mathcal{L}_\infty^n(D)$, respectively.

Definition 6 *The position $\mathbf{w}(\mathbf{x},t) = 0$ is defined to be **strongly asymptotically stable** if the displacement decays pointwise as follows*

$$\mathbf{w}(\mathbf{x},t) \to \mathbf{0} \text{ as } t \to \infty, \ \forall \mathbf{x} \in D, \tag{3.1}$$

for any $\mathbf{z}(\mathbf{x},0) \in \mathcal{Z}$.

Definition 7 *The position $\mathbf{w}(\mathbf{x},t) = 0$ is defined to be **strongly exponentially stable** if, for any $\mathbf{z}(\mathbf{x},0) \in \mathcal{Z}$, the displacement decays pointwise exponentially as follows*

$$\|\mathbf{w}(\mathbf{x},t)\|_e \le Ae^{-\lambda t}, \ \forall \mathbf{x} \in D, \ \forall t \in [0,\infty), \tag{3.2}$$

where $\|\cdot\|_e$ is the standard Euclidean norm in $\mathcal{R}^n$ and A, λ are positive constants.

In the control proofs, the generalized coordinates are assumed to be sufficiently smooth so that they and/or their time derivatives may be members of vector spaces $\mathcal{S}(D)$ and/or $\mathcal{S}_{\mathcal{A}}(D)$ defined as follows.

Definition 8 *If* $\mathbf{f}(\mathbf{x},t) \in \mathcal{S}^n(D)$, *then* $\|\mathbf{f}(\mathbf{x},t)\| \in \mathcal{L}_\infty$ *implies* $\mathbf{f}(\mathbf{x},t) \in \mathcal{L}_\infty^n(D)$.

Norm bounded elements of $\mathcal{S}(D)$ are pointwise bounded. Elements of $\mathcal{S}(D)$ do not exhibit delta function distributions that would cause an $L^2(D)$ bounded displacement to be pointwise unbounded.

Definition 9 *If* $\mathbf{f}(\mathbf{x},t) \in \mathcal{S}_{\mathcal{A}}^n(D)$, *then* $\mathbf{f}(\mathbf{x},t) \in \mathcal{L}_\infty^n(D)$ *implies* $\mathcal{A}\mathbf{f}(\mathbf{x},t) \in \mathcal{L}_\infty^n(D)$ *where* $\mathcal{A} : \mathcal{Z} \to D(\mathcal{A})$ *is an* $n \times n$ *matrix operator.*

Pointwise bounded elements of $\mathcal{S}_{\mathcal{A}}(D)$ are bounded if operated on by $\mathcal{A}$. Elements of $\mathcal{S}_{\mathcal{A}}(D)$ do not have discontinuities that would cause pointwise unbounded higher derivatives. The space $\mathcal{S}_{\mathcal{A}}(D)$ is more restrictive than $D(\mathcal{A})$ because $D(\mathcal{A})$ only requires that the highest derivatives in $\mathcal{A}$ be norm bounded where $\mathcal{S}_{\mathcal{A}}(D)$ requires pointwise boundedness.

3.2 Inequalities

Inequalities enable bounding of candidate Lyapunov functionals and their time derivatives. Thus, they are an important tool in Lyapunov analysis. Inequalities can be divided into algebraic and integral forms dealing with vector or function spaces, respectively. Norm based inequalities such as the triangle inequality and the Cauchy-Schwarz inequality can be applied to vector or function spaces to produce algebraic or integral inequalities, respectively [100]. The triangle inequality can be stated

$$\|\mathbf{w}_1 + \mathbf{w}_2\| \leq \|\mathbf{w}_1\| + \|\mathbf{w}_2\| \qquad \forall\, \mathbf{w}_1, \mathbf{w}_2 \in \mathcal{H}. \tag{3.3}$$

The Cauchy-Schwarz inequality is

$$|(\mathbf{w}_1, \mathbf{w}_2)| \leq \|\mathbf{w}_1\|\,\|\mathbf{w}_2\| \qquad \forall\, \mathbf{w}_1, \mathbf{w}_2 \in \mathcal{H}. \tag{3.4}$$

Lemma 10 *With* $\mathcal{H} = \{a\,|a \in H^1(0,L),\ a(0) = 0\}$, *if* $w(x,t) \in \mathcal{H}$ *then*

$$w^2(s,t) \leq L \int_0^L w_x^2\, dx \qquad \forall\, s \in [0,L]. \tag{3.5}$$

Proof: We define the inner product

$$(w_1, w_2) = \int_0^L w_1 w_2 dx, \tag{3.6}$$

and Cauchy-Schwarz Ineq. (3.4) is

$$\left[\int_0^L w_1 w_2 dx\right]^2 \le \int_0^L w_1^2 dx \int_0^L w_2^2 dx. \tag{3.7}$$

Substitution of $w_2 = w_x(x,t)$ and

$$w_1(x) = H(s-x) = \begin{cases} 1 & x \le s \\ 0 & x > s \end{cases}$$

into Eq. (3.7) yields

$$w^2(s,t) = (w(s,t) - w(0,t))^2 \le s \int_0^L w_x^2\, dx \le L \int_0^L w_x^2\, dx \quad \forall\, x \in [0,L] \tag{3.8}$$

due to the pinned boundary condition at $x = 0$ in $\mathcal{H}$.

□

Similarly, if $\mathcal{H} = \{a \,|a \in H^2(0,L),\, a_x(0) = 0\}$ with the inner product in Eq. (3.6), then

$$w_x^2(x,t) \le L \int_0^L w_{xx}^2\, dx \qquad \forall\, x \in [0,L]. \tag{3.9}$$

Lemma 11 *For $\mathcal{H} = \mathcal{R}$,*

$$\begin{array}{l} w_1 w_2 \le |w_1 w_2| \le w_1^2 + w_2^2 \\ 2w_1 w_2 \le 2\,|w_1 w_2| \le w_1^2 + w_2^2 \end{array} \qquad \forall w_1, w_2 \in \mathcal{H}. \tag{3.10}$$

Proof: For all $w_1, w_2 \in \mathcal{R}$ we have

$$(|w_1| - |w_2|)^2 = w_1^2 - 2\,|w_2|\,|w_1| + w_2^2 \ge 0. \tag{3.11}$$

Solution of Ineq. (3.11) yields

$$w_1^2 + w_2^2 \ge 2\,|w_2|\,|w_1| \ge 2w_1 w_2,$$

and

$$w_1^2 + w_2^2 \ge 2\,|w_2|\,|w_1| \ge |w_1 w_2| \ge w_1 w_2.$$

□

Ineq. (3.10) also produces the nonlinear damping inequality

$$|w_1 w_2| = \left|\left(\frac{1}{\sqrt{\delta}} w_1\right)\left(\sqrt{\delta} w_2\right)\right| \le \frac{1}{\delta} w_1^2 + \delta w_2^2 \qquad \forall w_1, w_2 \in \mathcal{R} \text{ and } \delta > 0. \tag{3.12}$$

Rayleigh's Quotient provides additional inequalities. If the operator $\mathcal{A}_0$ is positive and symmetric, then

$$(\mathcal{A}_0 \mathbf{w}, \mathbf{w}) \ge \lambda_{\min}(\mathcal{A}_0)\, \|\mathbf{w}\|^2, \tag{3.13}$$

where $\lambda_{\min}(\mathcal{A}_0)$ is the minimum eigenvalue of $\mathcal{A}_0$ given by solutions of $(\lambda\mathcal{I} - \mathcal{A}_0)\,\mathbf{w} = \mathbf{0}$ [120]. If the operator $\mathcal{A}_0$ is symmetric and bounded above, then

$$(\mathcal{A}_0\mathbf{w}, \mathbf{w}) \le \lambda_{\max}(\mathcal{A}_0)\, \|\mathbf{w}\|^2, \tag{3.14}$$

where $\lambda_{\max}(\mathcal{A}_0)$ is the maximum eigenvalue of $\mathcal{A}_0$. For $\mathbf{w} \in \mathcal{R}^n$, Ineqs. (3.13) and (3.14) produce

$$\lambda_{\min}(\mathbf{A}_0)\, \|\mathbf{w}\|_e^2 \le \mathbf{w}^T\mathbf{A}_0\mathbf{w} \le \lambda_{\max}(\mathbf{A}_0)\, \|\mathbf{w}\|_e^2, \tag{3.15}$$

where $\mathbf{A}_0$ is a symmetric, positive definite matrix.

Lemma 12 *For* $\mathcal{H} = \left\{[a, b]^T \,| a \in H^1(0, L),\ b \in \mathcal{R},\ a(0) = 0,\ b = a(L)\right\}$, *if* $\mathbf{w}(\mathbf{x}, t) = [w(x, t),\ w(L, t)]^T \in \mathcal{H}$ *then*

$$\int_0^L w_x^2 dx \ge \frac{1}{L^2}\int_0^L w^2 dx. \tag{3.16}$$

Proof: We define the inner product

$$(\mathbf{w}_1, \mathbf{w}_2) = \frac{1}{2}\int_0^L w_1 w_2 dx + \frac{1}{2\alpha} w_1(L) w_2(L) \tag{3.17}$$

and operator

$$\mathcal{A}_0\mathbf{w} = \begin{bmatrix} -w_{xx} \\ \alpha w_x(L, t) \end{bmatrix} \tag{3.18}$$

where $\alpha > 0$. The eigenfunctions of $\mathcal{A}_0$ have the following form

$$W(x) = \sin\beta x, \tag{3.19}$$

where the pinned boundary condition at $x = 0$ has been used and $\lambda = \beta^2$. The frequency equation is

$$\beta\sin\beta L - \alpha\cos\beta L = 0. \tag{3.20}$$

We are free to chose $\alpha = \frac{\sin(1)}{L\cos(1)}$ so the minimum solution of Eq. (3.20) is $\beta_1 = \frac{1}{L}$ and $\lambda_{\min} = \frac{1}{L^2}$. From Ineq. (3.13), we get

$$\begin{aligned}
(\mathcal{A}_0\mathbf{w}, \mathbf{w}) &= \frac{1}{2}\int_0^L -w_{xx} w dx + \frac{1}{2} w_x(L) w(L) \\
&= \frac{1}{2}\int_0^L w_x^2 dx \\
&\ge \lambda_{\min}(\mathcal{A}_0)\left(\frac{1}{2}\int_0^L w^2 dx + \frac{1}{2\alpha} w^2(L)\right) \\
&= \lambda_{\min}(\mathcal{A}_0)\, \|\mathbf{w}\|^2.
\end{aligned} \tag{3.21}$$

Substitution of $\lambda_{\min}$ into Ineq. (3.21) yields

$$\int_0^L w_x^2 dx \geq \frac{1}{L^2}\left(\int_0^L w^2 dx + \frac{1}{\alpha} w^2(L)\right) \geq \frac{1}{L^2}\int_0^L w^2 dx. \tag{3.22}$$

□

The tightest bound (largest β_1) in Eq. (3.20) occurs if we let $\alpha \to \infty$ ($\lambda_{\min} = \frac{\pi^2}{4L^2}$) as follows

$$\int_0^L w_x^2 dx \geq \frac{\pi^2}{4L^2}\int_0^L w^2 dx \tag{3.23}$$

but we often use the simpler form in Ineq. (3.16) for convenience. Similarly, if $w_x(0,t) = 0$, we have

$$\int_0^L w_{xx}^2 dx \geq \frac{\pi^2}{4L^2}\int_0^L w_x^2 dx > \frac{1}{L^2}\int_0^L w_x^2 dx. \tag{3.24}$$

3.3 Existence of a Semigroup Solution

A fundamental question in the study of distributed parameter systems is whether an unique solution exists to the partial differential equations of a particular model. Semigroup theory helps to answer these questions (See *e.g.* [5, 78, 99, 28, 115]). The purpose of this section is to demonstrate how we prove that a unique semigroup exists. This is an important first step in the application of LaSalle's Invariance Principle to prove asymptotic stability of a closed loop system. If we close the loop on the PDE/ODE models developed in Chapter 2 and eliminate all inputs, the state Eqs. (2.89) can be written

$$\dot{\mathbf{z}} = \mathcal{A}\mathbf{z}, \tag{3.25}$$

where $\mathbf{z}(t)$ is the state vector and $\mathcal{A}$ is the state matrix operator. We assume the model has an appropriately defined Hilbert Space $\mathbf{z}(t) \in \mathcal{Z}$ and initial conditions $\mathbf{z}_0 = \mathbf{z}(0)$.

Once the semigroup solution $T(t)$ of a system has been found, the response $\mathbf{z}(t)$ can be written in the simple form

$$\mathbf{z}(t) = T(t)\mathbf{z}_0. \tag{3.26}$$

Determination of the existence and uniqueness of a semigroup starts with the definition of an appropriate Hilbert Space $\mathcal{Z}$. Then, the operator $\mathcal{A}$ must be shown to be maximally dissipative and densely defined in $\mathcal{Z}$. Given these conditions, there exists a unique weak solution Eq. (3.26) for all $\mathbf{z}_0 \in \mathcal{Z}$, where $T(t)$ is the contraction semigroup generated by $\mathcal{A}$. Furthermore, if $\mathbf{z}_0 \in \mathcal{D}(\mathcal{A})$, then $\mathbf{z}(t) = T(t)\mathbf{z}_0$ is a strong solution.

3.3.1 Axially Moving String

We first consider the axially moving string model in Eqs. (2.27) - (2.31) with zero transport speed ($v = 0$), distributed damping ($c = 0$), distributed forcing ($f_u = 0$), and actuator mass ($m = 0$) given by

$$\rho w_{tt} - P w_{xx} = 0, \qquad \forall x \in (0, x_c) \cup (x_c, L), \tag{3.27}$$

where x_c is an irrational number between $(0, L)$ with boundary conditions

$$\begin{aligned} &w(0) = 0, \\ &w(L) = 0, \\ &w(x_c^-) = w(x_c^+), \\ &P w_x(x_c^-) - P w_x(x_c^+) = -k_d w_t(x_c, t), \end{aligned} \tag{3.28}$$

where a damping control law with feedback gain k_d has been used . The system energy

$$E = \frac{1}{2} \int_0^L (\rho w_t^2 + P w_x^2) dx, \tag{3.29}$$

is nonincreasing

$$\dot{E} = -k_d w_t^2(x_c, t) \leq 0. \tag{3.30}$$

We define the function space $\mathcal{Z}$ as follows,

$$\begin{aligned} \mathcal{Z} \;=\; & \left\{ \left[w^-, \nu^-, w^+, \nu^+ \right]^T | w^- \in H^1(0, x_c), \nu^- \in L^2(0, x_c), \right. \\ & w^+ \in H^1(x_c, L), \nu^+ \in L^2(x_c, L); w^-(0) = 0, w^+(L) = 0, \\ & \left. v^-(x_c) = v^+(x_c) = v(x_c), w^-(x_c) = w^+(x_c) = w(x_c) \right\}. \end{aligned} \tag{3.31}$$

The closed-loop system can be written in the following abstract form

$$\dot{\mathbf{z}} = \mathcal{A}\mathbf{z},$$

where $\mathbf{z} = [w^-, \nu^-, w^+, \nu^+]^T \in \mathcal{Z}$, $w = w(x,t)$, and $\nu = w_t(x,t)$. We define the operator $\mathcal{A} : \mathcal{D}(\mathcal{A}) \to \mathcal{Z}$ as

$$\mathcal{A}\mathbf{z} = \begin{bmatrix} \nu^- \\ \frac{1}{\rho} P w_{xx}^- \\ v^+ \\ \frac{1}{\rho} P w_{xx}^+ \end{bmatrix} \tag{3.32}$$

and the domain of $\mathcal{A}$ is

$$\begin{aligned} \mathcal{D}(\mathcal{A}) := \; & \left\{ [w^-, \nu^-, w^+, \nu^+]^T \in \mathcal{Z} | w^- \in H^2(0, x_c), \nu^- \in H^1(0, x_c), \right. \\ & \left. w^+ \in H^2(x_c, L), \nu^+ \in H^1(x_c, L); P w_x^-(x_c) - P w_x^+(x_c) = -k_d \nu(x_c) \right\}. \end{aligned}$$

We define the following inner product in $\mathcal{Z}$

$$\langle \mathbf{z}, \widetilde{\mathbf{z}} \rangle = \int_D \rho \nu \widetilde{\nu} dx + \int_D P w_x \widetilde{w}_x dx, \tag{3.33}$$

where $\widetilde{\mathbf{z}} = \left[\widetilde{w}^-, \widetilde{\nu}^-, \widetilde{w}^+, \widetilde{\nu}^+\right]^T \in \mathcal{Z}$ and $D = ((0, x_c) \bigcup (x_c, L))$.

It is straightforward to show that $\mathcal{Z}$ with the inner product given by Eq. (3.33) is a Hilbert space.

Lemma 13 *The operator $\mathcal{A}$ in Eq. (3.32) is a dissipative operator in $\mathcal{Z}$ defined in Eq. (3.31).*

Proof:

$$\begin{aligned} \langle \mathcal{A}\mathbf{z} - \mathcal{A}\widetilde{\mathbf{z}}, \mathbf{z} - \widetilde{\mathbf{z}} \rangle &= \int_D (P w_{xx} - P \widetilde{w}_{xx})(w_t - \widetilde{w}_t) dx \\ &+ \int_D P(w_{xt} - \widetilde{w}_{xt})(w_x - \widetilde{w}_x) dx \\ &= -k_d (v(x_c) - \widetilde{v}(x_c))^2 \leq 0. \end{aligned}$$

□

As the name implies, the dissipativity condition in the previous lemma is closely related to the time rate of change of energy in Eq. (3.30).

Lemma 14 *The operator $\mathcal{A}$ in Eq. (3.32) is maximal dissipative in $\mathcal{Z}$ defined in Eq. (3.31).*

Proof: The operator is maximal dissipative if it is dissipative (See Lemma 13) and for $\lambda > 0$ and any $\widetilde{\mathbf{z}} \in \mathcal{Z}$, there exists a $\mathbf{z} \in \mathcal{D}(\mathcal{A})$ such that $\lambda \mathbf{z} - \mathcal{A}\mathbf{z} = \widetilde{\mathbf{z}}$. Using the operator definition (3.32), we obtain the following equations for $x \in D$

$$\left[\begin{array}{l} \lambda w - \nu = \widetilde{w} \\ \lambda \nu - \frac{1}{\rho} P w_{xx} - \widetilde{\nu} \end{array} \right]. \tag{3.34}$$

Combination of Eqs. (3.34) produces

$$\lambda^2 w - \frac{1}{\rho} P w_{xx} = \lambda \widetilde{w} + \widetilde{\nu}, \text{ for all } x \in D. \tag{3.35}$$

The solution of Eq. (3.35) satisfying $w(0) = 0$ and $w(L) = 0$ is given by

$$\begin{aligned} w^-(x) &= c_1 \sinh(\sqrt{\frac{\rho}{P}} \lambda x) - \sqrt{\frac{\rho}{P}} \frac{1}{\lambda} \int_0^{x_c^-} (\lambda \widetilde{w}(s) + \widetilde{\nu}(s)) \sinh(\sqrt{\frac{\rho}{P}} \lambda (x - s)) ds, \\ w^+(x) &= c_2 \sinh(\sqrt{\frac{\rho}{P}} \lambda (L - x)) \\ &- \sqrt{\frac{\rho}{P}} \frac{1}{\lambda} \int_0^{L - x_c^+} (\lambda \widetilde{w}(s) + \widetilde{\nu}(s)) \sinh(\sqrt{\frac{\rho}{P}} \lambda (L - x - s)) ds \end{aligned}$$

where c_1 and c_2 are uniquely determined by $w^-(x_c) = w^+(x_c)$ and $Pw_x^-(x_c) - Pw_x^+(x_c) = -k_d\nu(x_c) = -k_d(\lambda w(x_c) - \widetilde{w}(x_c))$. The solution $w \in H^2$ because it involves integration of $\widetilde{w} \in H^1$ and $v \in H^1$ from $\nu = \lambda w - \widetilde{w}$. Thus, $\mathbf{z} \in \mathcal{D}(\mathcal{A})$.
□

Lemma 15 *The operator $\mathcal{A}$ in Eq. (3.32) is densely defined in $\mathcal{Z}$ defined in Eq. (3.31).*

Proof: As in [37], we first note that given any $h \in L^2(0, L)$ and $c \in R$, for any $\varepsilon > 0$, there must be $h_1 \in C^\infty(0, L)$, the set of all infinitely differentiable functions on $(0, L)$, such that

$$\|h - h_1\|_2 < \varepsilon.$$

We must prove that for any fixed $\widetilde{\mathbf{z}} \in \mathcal{Z}$ and any $\varepsilon > 0$, there exists a $\mathbf{z} \in \mathcal{D}(\mathcal{A})$ such that $\|\widetilde{\mathbf{z}} - \mathbf{z}\| < \varepsilon$, where $\|\mathbf{z}\| = \sqrt{\langle \mathbf{z}, \mathbf{z}\rangle}$ and $\langle \mathbf{z}, \mathbf{z}\rangle$ is defined in Eq. (3.33). Since $\widetilde{\nu} \in L^2(0, L)$, there exists $\nu \in C^\infty(0, L)$ such that

$$\int_0^L \rho(\widetilde{\nu} - \nu)^2 dx < \varepsilon^2/2, \tag{3.36}$$

Similarly, $\widetilde{w} \in H^1(D)$ so $\widetilde{w}_x \in L^2(D)$, and there exists $w_x \in C^\infty(D)$ such that

$$\int_D P(\tilde{w}_x - w_x)^2 dx < \varepsilon^2/2, \tag{3.37}$$

where we choose $w^-(0) = 0$, $w^-(x_c) = w^+(x_c)$, $w^+(L) = 0$, and $w_x^-(x_c) - w_x^+(x_c) = -k_d\nu(x_c)/P$. Denoting $\mathbf{z} = [w, \nu]^T$, we have $w, \nu \in C^\infty(D)$. Thus, combining Eqs. (3.36) and (3.37), we obtain $\|\widetilde{\mathbf{z}} - \mathbf{z}\|_{\mathcal{Z}} < \varepsilon$ with $\mathbf{z} \in \mathcal{D}(\mathcal{A})$.
□

Theorem 16 *The operator $\mathcal{A}$ in Eq. (3.32) generates a strongly continuous C_0-semigroup of contraction $T(t)$ on $\mathcal{Z}$ defined in Eq. (3.31).*

Proof: The maximal dissipative and densely defined properties of $\mathcal{A}$ proven in Lemmas 18 and 19 mean that $\mathcal{A}$ generates a strongly continuous C_0-semigroup of contraction $T(t)$ on $\mathcal{Z}$ [99].
□

3.3.2 Gantry Crane

We now prove that the gantry crane Eqs. (2.21) - (2.23) have an unique solution. We define the function space $\mathcal{Z}$ as follows

$$\begin{aligned}\mathcal{Z} := \Big\{ & [w, \nu, \varphi, \psi, \vartheta, \varrho]^T \mid w \in H^1(0,1), \nu \in L^2(0,1), \varphi, \psi, \vartheta, \varrho \in R; \\ & \varphi = w(0), \psi = \nu(0), \vartheta = w(1), \varrho = \nu(1) \big\}.\end{aligned} \tag{3.38}$$

The closed loop system given by Eqs. (2.100), (2.101), and (2.102) can be written as an abstract state Eq. (3.25) where $\mathbf{z} = [w, \nu, \varphi, \psi, \vartheta, \varrho]^T \in \mathcal{Z}$, $w = w(x,t)$, $\nu = w_t(x,t)$, $\varphi = w(0,t)$, $\psi = w_t(0,t)$, $\vartheta = w(1,t)$, and $\varrho = w_t(1,t)$. The operator $\mathcal{A} : \mathcal{D}(\mathcal{A}) \to \mathcal{Z}$ is an unbounded operator defined as

$$\mathcal{A}\begin{bmatrix} w \\ \nu \\ \varphi \\ \psi \\ \vartheta \\ \varrho \end{bmatrix} = \begin{bmatrix} \nu \\ ((p_o - x)\, w_x)_x \\ \psi \\ -k_p\varphi - k_d\psi + k_a w_x(0) \\ \varrho \\ -w_x(1) \end{bmatrix}, \tag{3.39}$$

where the domain of $\mathcal{A}$ is given by

$$\begin{aligned}\mathcal{D}(\mathcal{A}) := \; & \left\{ [w, \nu, \varphi, \psi, \vartheta, \varrho]^T \in \mathcal{Z} \mid w \in H^2(0,1), \nu \in H^1(0,1), \varphi, \psi, \vartheta, \varrho \in R; \right. \\ & \left. \varphi = w(0), \psi = \nu(0), \vartheta = w(1), \varrho = \nu(1) \right\}.\end{aligned} \tag{3.40}$$

We define the following inner product in $\mathcal{Z}$

$$\langle \mathbf{z}, \widetilde{\mathbf{z}} \rangle = \int_0^1 (p_o - x)\nu\widetilde{\nu}dx + \int_0^1 (p_o - x)^2 w_x \widetilde{w}_x dx + \frac{p_o^2}{k_a}\psi\widetilde{\psi} + (p_o - 1)^2 \varrho\widetilde{\varrho} + \frac{p_o^2 k_p}{k_a}\varphi\widetilde{\varphi}, \tag{3.41}$$

where $\mathbf{z} = [w, \nu, \varphi, \psi, \vartheta, \varrho]^T \in \mathcal{Z}$ and $\widetilde{\mathbf{z}} = \left[\widetilde{w}, \widetilde{\nu}, \widetilde{\varphi}, \widetilde{\psi}, \widetilde{\vartheta}, \widetilde{\varrho}\right]^T \in \mathcal{Z}$. It can be shown that $\mathcal{Z}$, with the inner product Eq. (3.41), is a Hilbert space.

Lemma 17 *The operator $\mathcal{A}$ in Eq. (3.39) is a dissipative operator in $\mathcal{Z}$ defined in Eq. (3.38).*

Proof:

$$\langle \mathcal{A}\mathbf{z} - \mathcal{A}\widetilde{\mathbf{z}}, \mathbf{z} - \widetilde{\mathbf{z}} \rangle = -\frac{p_o^2 k_d}{k_a}\left(\psi - \widetilde{\psi}\right)^2 \leq 0. \tag{3.42}$$

□

Lemma 18 *The operator $\mathcal{A}$ in Eq. (3.39) is maximal dissipative in $\mathcal{Z}$ defined in Eq. (3.38).*

Proof: Maximal dissipativity requires that $\mathcal{A}$ be dissipative and that the inverse of $\lambda\mathcal{I} - \mathcal{A}$ exists for some $\lambda > 0$. Thus, we show that for any $\widetilde{\mathbf{z}} = [\widetilde{w}, \widetilde{\nu}, \widetilde{\varphi}, \widetilde{\psi}, \widetilde{\vartheta}, \widetilde{\varrho}]^T \in \mathcal{Z}$, there exists a unique $\mathbf{z} = [w, \nu, \varphi, \psi, \vartheta, \varrho]^T \in \mathcal{D}(\mathcal{A})$ such that $\lambda\mathbf{z} - \mathcal{A}\mathbf{z} = \widetilde{\mathbf{z}}$, or

$$\begin{aligned} & \lambda w - \nu = \widetilde{w}, \\ & \lambda\nu - ((p_o - x)\, w_x)_x = \widetilde{\nu}, \\ & \lambda\varphi - \psi = \widetilde{\varphi}, \\ & \lambda\psi - (-k_p\varphi - k_d\psi + k_a w_x(0)) = \widetilde{\psi}, \\ & \lambda\vartheta - \varrho = \widetilde{\vartheta}, \\ & \lambda\varrho + w_x(1) = \widetilde{\varrho}. \end{aligned} \tag{3.43}$$

Elimination of ν, ψ, and ϱ from Eq. (3.43) produces

$$\begin{aligned} &\lambda^2 w - ((p_o - x)\, w_x)_x = \lambda \widetilde{w} + \widetilde{\nu}, \\ &\lambda (\lambda + k_d)\, w(0) - (-k_p w(0) + k_a w_x(0)) = (\lambda + k_d)\, \widetilde{\varphi} + \widetilde{\psi}, \\ &\lambda^2 w(1) + w_x(1) = \lambda \widetilde{\vartheta} + \widetilde{\varrho}. \end{aligned} \tag{3.44}$$

The Sturm-Liouville problem Eq. (3.44) admits a unique solution in $w \in H^2([0,1])$ if we choose λ different from the eigenvalues of the homogeneous Sturm-Liouville problem (*i.e.*, right hand side of Eqs. (3.44) equal to zero) (see p.277 [9]). Then ν can be determined from $\nu = \lambda w - \widetilde{w} \in H^1(0,1)$, so $\mathbf{z} = [w, \nu, \varphi, \psi, \vartheta, \varrho]^T \in \mathcal{D}(\mathcal{A})$.
□

Lemma 19 *The operator $\mathcal{A}$ in Eq. (3.39) is densely defined in $\mathcal{Z}$ defined in Eq. (3.38).*

Proof: As in [37], we first note that given any $h \in L^2(0,1)$ and $c_0, c_1 \in \mathcal{R}$, for any $\varepsilon > 0$, there exists an $h_1 \in C^\infty(0,1)$ such that

$$h_1(0) = c_0, \qquad h_1(1) = c_1, \qquad \text{and} \qquad \|h - h_1\|_2 < \varepsilon.$$

We now prove that for any $\widetilde{\mathbf{z}} = [\widetilde{w}, \widetilde{\nu}, \widetilde{\varphi}, \widetilde{\psi}, \widetilde{\vartheta}, \widetilde{\varrho}]^T \in \mathcal{Z}$ and any $\varepsilon > 0$, there exists a $\mathbf{z} = [w, \nu, \varphi, \psi, \vartheta, \varrho]^T \in \mathcal{D}(\mathcal{A})$ such that $\|\widetilde{\mathbf{z}} - \mathbf{z}\|_{\mathcal{Z}} < \varepsilon$, where $\|\mathbf{z}\|_{\mathcal{Z}} = \sqrt{\langle \mathbf{z}, \mathbf{z} \rangle}$. We first choose $a, b, c \in \mathcal{R}$ such that

$$\frac{p_o^2 k_p}{k_a} (\widetilde{\varphi} - a)^2 < \varepsilon^2/5, \qquad \frac{p_o^2}{k_a} (\widetilde{\psi} - b)^2 < \varepsilon^2/5, \qquad (p_o - 1)^2 (\widetilde{\varrho} - c)^2 < \varepsilon^2/5.$$

Since $\widetilde{w} \in H^1(0,1)$ and $\widetilde{\nu} \in L^2(0,1)$, there exists $\hat{w}, \hat{\nu} \in C^\infty(0,1)$, such that $\hat{w}(0) = a$, $\hat{\nu}(0) = b$, $\hat{\nu}(1) = c$ and

$$\int_0^1 (p_o - x)^2 (\widetilde{w}_x - \hat{w}_x)^2 dx < \varepsilon^2/5, \qquad \int_0^1 (\widetilde{\nu} - \hat{\nu})^2 dx < \varepsilon^2/5.$$

Denoting $\mathbf{z} = [\hat{w}, \hat{\nu}, a, b, \hat{w}(1), c]^T \in \mathcal{D}(\mathcal{A})$ produces

$$\begin{aligned} \|\widetilde{\mathbf{z}} - \mathbf{z}\|_{\mathcal{Z}} = &\left(\int_0^1 (p_o - x) (\widetilde{\nu} - \hat{\nu})^2 dx + \int_0^1 (p_o - x)^2 (\widetilde{w}_x - \hat{w}_x)^2 dx + \tfrac{p_o^2}{k_a} (\widetilde{\psi} - b)^2 \right. \\ &\left. + (p_o - 1)^2 (\widetilde{\varrho} - c)^2 + \tfrac{p_o^2 k_p}{k_a} (\widetilde{\varphi} - a)^2 \right)^{\frac{1}{2}} < \varepsilon, \end{aligned} \tag{3.45}$$

implying that $\mathcal{D}(\mathcal{A})$ is dense in $\mathcal{Z}$.
□

The following theorem follows directly from semigroup theory [99].

Theorem 20 *The closed loop system given by Eq. (3.25) possesses the unique weak solution $\mathbf{z}(t) = T(t)\mathbf{z}_0$ for $\mathbf{z}_0 \in \mathcal{Z}$, where $T(t)$ is the contraction semigroup generated by $\mathcal{A}$ defined in Eq. (3.39). Furthermore, if $\mathbf{z}_0 \in \mathcal{D}(\mathcal{A})$, then $\mathbf{z}(t) = T(t)\mathbf{z}_0$ is a strong solution.*

3.4 Stability Theorems

3.4.1 Lyapunov's Direct Method

We now present the main results of Lyapunov's direct method. For further details, interested readers are referred to [79, 49, 91, 72].

Theorem 21 *The position* $\mathbf{w}(\mathbf{x},t) = 0$ *is strongly stable if there exists a functional* $V(\mathbf{w})$ *such that* $\beta \|\mathbf{w}\|_e^2 \geq V(\mathbf{w}) \geq \alpha \|\mathbf{w}\|_e^2$ *for some constant* $\alpha, \beta > 0$ *and* $\dot{V} \leq 0$ *for all* $t \geq 0$ *and* $\mathbf{x} \in D$. *The position* $\mathbf{w}(\mathbf{x},t) = 0$ *is strongly asymptotically stable if, in addition,* $V \to 0$ *as* $t \to \infty$. *The position* $\mathbf{w}(\mathbf{x},t) = 0$ *is strongly exponentially stable if, in addition to the requirements for stability,* $\dot{V} \leq -\lambda \|\mathbf{w}\|_e^2$ *for all* $\mathbf{x} \in D$.

The conditions for exponential stability imply that

$$\dot{V} \leq -\lambda \|\mathbf{w}\|_e^2 \leq -\frac{\lambda}{\beta} V = -\gamma V \tag{3.46}$$

with the maximal solution

$$V(t) \leq V(0) e^{-\gamma t}. \tag{3.47}$$

so

$$\|\mathbf{w}(\mathbf{x},t)\|_e \leq \sqrt{\frac{\beta \left| w(\mathbf{x},0) \right|^2}{\alpha} e^{-\gamma t}}$$

and $\mathbf{w}(\mathbf{x},t)$ decays to zero exponentially fast for all $\mathbf{x} \in D$.

Lemma 22 *Given the positive scalar functional* $V(\mathbf{w})$ *with*

$$\dot{V} \leq -\gamma V + \epsilon, \tag{3.48}$$

and $\gamma, \epsilon > 0$,

$$V \leq V(0) e^{-\gamma t} + \frac{\epsilon}{\gamma}\left(1 - e^{-\gamma t}\right), \tag{3.49}$$

for all $t \in [0, \infty)$.

Proof: The result in Ineq. (3.49) is the maximal solution of Ineq. (3.48).
□

3.4.2 Barbalat's Lemma

For systems with a negative semidefinite Lyapunov functional time derivative, Barbalat's Lemma [118] can often provide an asymptotic result.

Theorem 23 *If* $V(t)$ *is a positive scalar function lower bounded by zero,* $\dot{V}(t) \leq -f(t)$ *where* $f(t)$ *is a positive scalar function, and* $\dot{f} \in L_\infty$, *then*

$$\lim_{t \to \infty} f(t) = 0. \tag{3.50}$$

Proof: See [118].
□

3.4.3 LaSalle's Invariance Principle

LaSalle's Invariance Principle extended to infinite dimensional systems (see [29] for examples) applies to systems governed by abstract evolution equations $\dot{\mathbf{z}} = \mathcal{A}\mathbf{z}$ with unique solutions. Often we have cases where the energy $E(t) = \frac{1}{2}\langle \mathbf{z}, \mathbf{z} \rangle > 0$ has a negative time derivative $\dot{E} \leq 0$. Thus, asymptotic stability cannot be proven using standard Lyapunov arguments. LaSalle's invariance principle states that all solutions of the system asymptotically tend to the maximal invariant subset of the following set:

$$\mathcal{S} = \left\{ \mathbf{z} \in \mathcal{Z} \mid \dot{E} = 0 \right\}, \tag{3.51}$$

provided that the solution trajectories for $t \geq 0$ are precompact in $\mathcal{Z}$. Precompactness of solution trajectories is proven by showing that for the operator $\mathcal{A}$, $0 \in \mathcal{R}(\mathcal{A})$ and $(\lambda\mathcal{I} - \mathcal{A})^{-1}$ is compact for some $\lambda > 0$.

Theorem 24 *Given (i) the abstract evolution equation $\dot{\mathbf{z}} = \mathcal{A}\mathbf{z}$ where $\mathcal{A}$ generates a strongly continuous C_0-semigroup of contraction $T(t)$ on $\mathcal{Z}$; (ii) the energy $E(t) = \frac{1}{2}\langle \mathbf{z}, \mathbf{z} \rangle > 0$ has a negative time derivative $\dot{E} \leq 0$; and (iii) all solution trajectories for $t \geq 0$ are precompact in $\mathcal{Z}$. If, for initial data $\mathbf{z}_0 \in \mathcal{Z}$, the condition $\dot{E}(t) = 0, \forall t \geq 0$ ($\dot{E}(t) \equiv 0$) implies $\mathbf{z}_0 = \mathbf{0}$, then the origin $\mathbf{z} = \mathbf{0}$ is asymptotically stable.*

Chapter 4
PASSIVE CONTROL

Although most vibration and noise control systems are open loop stable, the natural damping in these systems can be quite small. Thus, the first objective of a feedback control system is to ensure closed loop stability. To improve on open loop performance, asymptotic stabilization requires controllers that dissipate energy and drive the system response to zero. Under forcing, undamped systems may resonate, resulting in unbounded response. Natural or feedback damping can eliminate resonances and guarantee bounded response for bounded forcing. This chapter applies the mathematical tools introduced in Chapter 3 to the development of passive control systems for distributed vibration and noise. Passive controllers are defined as control laws that are not explicitly based on the system model.

4.1 Damping

In the absence of disturbances, the inherent damping in mechanical systems causes vibration energy to decay over time. Damping eliminates the infinite response associated with resonant forcing. Damping results from viscoelastic behavior in the material, frictional interaction between contacting surfaces, or movement through a dissipative fluid. Many practical control designs rely on sufficient damping to stabilize high frequency or unmodeled dynamics. Damping is difficult to quantify and may change over time, however, making robust performance questionable. In this section, the stability characteristics of systems with boundary, distributed, pointwise, and parametric damping are studied.

4.1.1 Boundary Damping

The pinned-controlled string and boundary damped cantilevered beam are used to demonstrate the effects of damping for second and fourth order systems, respectively. Similar results can be developed for other second order models (*e.g.* cranes, cables, axially moving webs, and ducts) and fourth order models such as the axially moving beam and flexible link robot arm.

Boundary Damped String

If we neglect the actuator inertia and substitute a boundary damper for the control force in Eqs. (2.14) - (2.16), we obtain the boundary-damped, one-dimensional wave equation model

$$\rho w_{tt} - P w_{xx} = 0 \text{ for } x \in (0, L), \quad (4.1)$$

$$w(0,t) = 0, \quad (4.2)$$

$$P w_x(L,t) = -k_d w_t(L,t). \quad (4.3)$$

where k_d is the boundary damping gain. The generalized coordinate $\mathbf{w}(\mathbf{x}, t) = w(x,t)$ with

$$\ddot{w} = -\mathcal{A}_0 \mathbf{w} = \frac{P}{\rho} w_{xx}. \quad (4.4)$$

The string energy

$$E = \Pi_k + \Pi_p = \frac{1}{2} \int_0^L \left(\rho w_t^2 + P w_x^2 \right) dx \quad (4.5)$$

is positive, making it a suitable Lyapunov functional candidate. The time derivative of Eq. (4.5)

$$\dot{E} = \int_0^L \left(\rho w_t w_{tt} + P w_x w_{xt} \right) dx \quad (4.6)$$

can be simplified using the field Eq. (4.1) and integration by parts as follows

$$\int_0^L \rho w_{tt} w_t dx = \int_0^L P w_{xx} w_t dx = P w_x w_t |_0^L - \int_0^L P w_x w_{xt} dx. \quad (4.7)$$

The second term in Eq. (4.7) cancels the second term in Eq. (4.6). The boundary condition in Eq. (4.2) implies $w_t(0,t) = 0$, so Eqs. (4.7) and (4.6) simplify to

$$\dot{E} = P w_x(L,t) w_t(L,t) = -k_d w_t^2(L,t) \leq 0, \quad (4.8)$$

where the boundary condition Eq. (4.3) has been used.

The result in Eq. (4.8) shows that $E < \infty$ or the system is weakly stable. Unfortunately, the velocity dependent time derivative in Eq. (4.8) contains no displacement related terms (*e.g.* $w_x(x,t)$) so it cannot be bounded by the system energy. Hence, an asymptotic or exponential stability result cannot be obtained using Lyapunov theory alone. We can use the Ineq. (3.8) and boundary condition Eq. (4.2) to show

$$w^2(x,t) \leq L \int_0^L w_x^2 \, dx \leq \frac{2L}{P} E < \infty \quad \forall \, x \in [0, L], \quad (4.9)$$

proving $w(x,t) \in \mathcal{L}_\infty(0,L)$ and pointwise or strong stability of the zero position.

An important part of control development is to check the boundedness of all signals in closed loop operation. Unstable pole/zero cancellation, for example, causes unstable internal dynamics despite an asymptotically stable output response. For the boundary damped string, we can show that the displacement is bounded. If we assume $w_t(x,t) \in \mathcal{S}(0,L)$ then $\|w_t\|_2 < \sqrt{\frac{2}{\rho} E} < \infty$ implies $w_t(x,t) \in \mathcal{L}_\infty(0,L)$ where

$\|\cdot\|_2$ is the standard L_2 norm. Thus the boundary damping force $k_d w_t(L,t) \in \mathcal{L}_\infty$ and, from Eq. (4.3), $w_x(L,t) \in \mathcal{L}_\infty$. If we assume $w(x,t) \in \mathcal{S}_{A_0}(0,L)$, then Eq. (4.9) implies that $w_{xx}(x,t) \in \mathcal{L}_\infty(0,L)$ and the field Eq. (4.1) yields $w_{tt}(x,t) \in \mathcal{L}_\infty(0,L)$. Thus all closed loop signals are bounded.

Barbalat's Lemma or LaSalle's Invariance Principle may be able to provide a stronger stability result. Barbalat's lemma requires the calculation of

$$\ddot{E} = -2k_d w_{tt}(L,t) w_t(L,t). \tag{4.10}$$

Based on the previous assumptions, $\ddot{E} \in \mathcal{L}_\infty$. Thus, we know that

$$\dot{E} = -k_d w_t^2(L,t) \to 0 \text{ as } t \to \infty$$

so the boundary velocity is asymptotically stabilized. This is stronger than the previous result but insufficient to prove asymptotic stability of the generalized coordinate $w(x,t)$.

If we assume that the closed loop system given by Eqs. (4.1) - (4.3) generates a strongly continuous C_0-semigroup of contraction $T(t)$ on $\mathcal{H}$, we can apply LaSalle's invariance principle (Theorem 24). We need to show that $\dot{E}(t) \equiv 0$ implies $w(x,t) = 0$. If $\dot{E}(t) \equiv 0$ then $w_t(L,t) \equiv 0$ and, from Eq. (4.3), $w_x(L,t) \equiv 0$. Thus, the system response is governed by the equations

$$\begin{aligned} &\rho w_{tt} - P w_{xx} = 0 \text{ for } x \in (0,L), && \text{(a)} \\ &w(0,t) = 0, && \text{(b)} \\ &w_x(L,t) = 0, && \text{(c)} \\ &w_t(L,t) = 0. && \text{(d)} \end{aligned} \tag{4.11}$$

Since all the boundary conditions are separable, the solution of Eqs. (4.11) can be found by using separation of variables [81] as follows:

$$w(x,t) = W(x)\eta(t). \tag{4.12}$$

We distinguish two cases that produce $w_t(L,t) \equiv 0$:

i) $\dot{\eta}(t) \equiv 0$. From Eq. (4.11a), it follows that $PW_x(x) = c_1$, where c_1 is an arbitrary constant. From Eq. (4.11c), however, we obtain $c_1 = 0$, so $W(x) = c_2$. Eq. (4.11b) requires $c_2 = 0$, yielding $w(x,t) = 0$.

ii) $W(L) = 0$. The eigenvalue problem generated by substitution of Eq. (4.12) into (4.11a) with $\eta(t) = e^{i\omega t}$ is

$$\omega^2 \rho W + P W_{xx} = 0. \tag{4.13}$$

The solutions of Eq. 4.13 that satisfy Eq. (4.11b) are $W(x) = c \sin \beta x$ where $\beta = \frac{\rho\omega^2}{P}$. From Eqs. (4.11c) and (4.11d) we get

$$c \cos \beta L = 0, \tag{4.14}$$

$$c \sin \beta L = 0, \tag{4.15}$$

where, for nonzero $W(x)$, we have $\beta \neq 0$. Squaring and adding Eqs. (4.14) - (4.15) produces $c^2(\cos^2\beta L + \sin^2\beta L) = 0$, yielding $c = 0$ and $w(x,t) = 0$. Therefore the closed loop displacement is asymptotically stable.

The strongest stability result (exponential) for the boundary damped string results from using Lyapunov's method with a Lyapunov functional other than the system energy.

Theorem 25 *The boundary-damped string given by Eqs. (4.1) - (4.3) is strongly exponentially stable.*

Proof: We define a new Lyapunov functional

$$V(t) = E(t) + C(t), \tag{4.16}$$

where $E(t)$ is defined in Eq. (4.5) with

$$\frac{1}{2}\min(\rho, P)\int_0^L \left(w_x^2 + w_t^2\right) dx \le E(t) \le \frac{1}{2}\max(\rho, P)\int_0^L \left(w_x^2 + w_t^2\right) dx \tag{4.17}$$

and

$$C(t) = \beta\int_0^L x w_x w_t dx, \tag{4.18}$$

where β is a positive constant. First, we check to see if $V > 0$. The crossing term

$$|C(t)| = \beta\left|\int_0^L x w_x w_t dx\right| \le \beta\int_0^L x\left(w_x^2 + w_t^2\right) dx \le \beta L\int_0^L \left(w_x^2 + w_t^2\right) dx, \tag{4.19}$$

where Ineqs. (3.10) and (4.17) and $L \ge x$ have been used. Comparing Eq. (4.19) with Eq. (4.5), we obtain

$$|C(t)| \le \frac{2\beta L}{\min(\rho, P)}E(t),$$

or

$$-\frac{2\beta L}{\min(\rho, P)}E(t) \le C(t) \le \frac{2\beta L}{\min(\rho, P)}E(t). \tag{4.20}$$

Thus, the Lyapunov functional in Eq. (4.16) is bounded as follows:

$$\lambda_1 E(t) \le V(t) \le \lambda_2 E(t), \tag{4.21}$$

where

$$\begin{aligned}
\lambda_1 &= 1 - \frac{2\beta L}{\min(\rho, P)} > 0, \\
\lambda_2 &= 1 + \frac{2\beta L}{\min(\rho, P)} > 1,
\end{aligned}$$

provided

$$0 < \beta < \frac{\min(\rho, P)}{2L}. \tag{4.22}$$

The Lyapunov functional time derivative

$$\dot{V} = \dot{E} + \dot{C} \tag{4.23}$$

involves the energy time derivative in Eq. (4.8) and

$$\dot{C} = \dot{C}_1 + \dot{C}_2, \tag{4.24}$$

where

$$\dot{C}_1 = \beta \int_0^L x w_x w_{tt} dx, \tag{4.25}$$

$$\dot{C}_2 = \beta \int_0^L x w_{xt} w_t dx. \tag{4.26}$$

Substitution of the field and boundary conditions Eqs. (4.1) - (4.3) and energy derivative Eq. (4.8) into Eq. (4.25) yields

$$\begin{aligned}
\dot{C}_1 &= \beta \int_0^L x w_x \frac{P}{\rho} w_{xx} dx = \frac{\beta P}{\rho}\left(x w_x^2\Big|_0^L - \int_0^L \left[x w_x\right]_x w_x dx \right) \\
&= \frac{\beta P}{\rho}\left(L w_x^2(L,t) - \int_0^L w_x^2 dx \right) - \beta \int_0^L x w_x w_{tt} dx \\
&= \frac{\beta P}{\rho}\left(\frac{L k_d^2}{P^2} w_t^2(L,t) - \int_0^L w_x^2 dx \right) - \dot{C}_1,
\end{aligned} \tag{4.27}$$

where integration by parts has been used. Solution of Ineq. (4.27) results in

$$\dot{C}_1 = \frac{\beta P}{2\rho}\left(\frac{L k_d^2}{P^2} w_t^2(L,t) - \int_0^L w_x^2 dx \right). \tag{4.28}$$

Integration by parts of Eq. (4.26) gives

$$\begin{aligned}
\dot{C}_2 &= \beta \left(x w_t^2\Big|_0^L - \int_0^L w_t^2 + x w_{xt} w_t dx \right) \\
&= \beta \left(L w_t^2(L,t) - \int_0^L w_t^2 dx \right) - \dot{C}_2.
\end{aligned} \tag{4.29}$$

Solution of Eq. (4.29) yields

$$\dot{C}_2 = \frac{\beta}{2}\left(L w_t^2(L,t) - \int_0^L w_t^2 dx \right). \tag{4.30}$$

Now we combine Eqs. (4.8), (4.28), and (4.30) to produce

$$\begin{aligned}\dot{V} &= -\left(k_d - \frac{\beta L k_d^2}{2\rho P} - \frac{\beta L}{2}\right) w_t^2(L,t) - \frac{\beta}{2}\left(\int_0^L w_x^2 dx + \int_0^L w_t^2 dx\right) \\ &\leq -\lambda_3 E \leq -\frac{\lambda_3}{\lambda_2} V,\end{aligned} \tag{4.31}$$

where Eq. (4.21) has been used,

$$\lambda_3 = \frac{\beta}{\max(\rho, P)},$$

and

$$k_d - \frac{\beta L k_d^2}{2\rho P} - \frac{\beta L}{2} > 0$$

or

$$\beta < \frac{2\rho P k_d}{L\left(k_d^2 + \rho P\right)}. \tag{4.32}$$

The maximal solution of Ineq. (4.31) is given by Ineq. (3.47). Ineqs. (4.9) and (4.21) produces the desired result

$$w^2(x,t) \leq \frac{2L}{P} E \leq \frac{2L}{\lambda_1 P} V(0) e^{\left(-\frac{\lambda_3}{\lambda_2} t\right)}.$$

□

Experimental Setup Now we implement the boundary damper using a mechatronic system comprising a mechanical system, interface electronics, and real-time software. A schematic diagram of the experimental setup is shown in Fig. 1. The mechanical setup consists of an extensible string pinned at one end and attached to a linearly translating gantry at the other end. The gantry is driven by a brushless DC motor (Baldor model 3300) via a belt-pulley transmission. The gantry rides on two parallel 1-inch diameter steel rods on linear bearings. The mechanical parameter values are $m = 3.5$ kg, $\rho = 0.03$ kg/m, and $L = 0.31$ m. The displacement of the gantry, $w(L,t)$, is obtained from a 1000-count rotary encoder (Hohner) attached to the motor shaft. The static and dynamic tension measurement of the string, $P(t)$, is obtained from the JR3 force sensor attached to the string at the pinned end. For the linear model presented in this section, $P(t) = P$ constant and this sensor is not used. In Chapter 5 we will develop an active controller for the nonlinear model that requires tension measurement. A hollow-shaft 1000-count rotary encoder is mounted on the gantry to measure the string deflection angle, $w_x(L,t)$ at the free end.

A Pentium 166 MHz PC running QNX (a real-time micro-kernel based operating system) hosts the control algorithm. *Qmotor* [104], a graphical user-interface

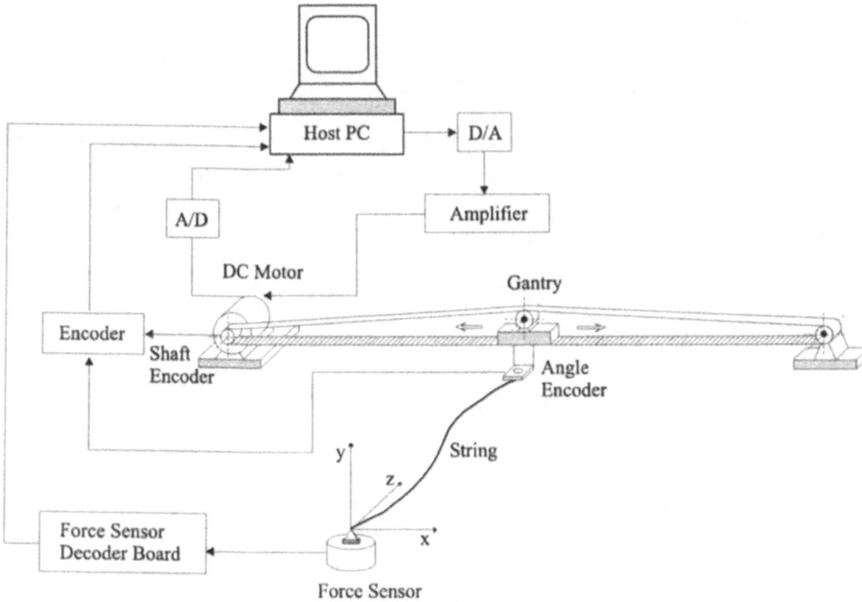

Figure 1 Boundary controlled string experimental setup.

developed in-house, provides an environment to write the control algorithm in the 'C' programming language. It also provides features such as on-line graphing and allows the user to vary control gains without having to recompile the program. A sample time of 0.5 ms is used. The *MultiQ* I/O board provides for data transfer between the computer subsystem and the electrical interface. An A/D channel is used to measure the current flowing through the windings of the DC motor sensed by a Hall-effect current sensor. One D/A channel is used to output a voltage to drive the DC motor. This voltage is first amplified by an OP07 precision operational amplifier circuit and then by a Techron linear power amplifier that sources up to current of 10 amps at 100 volts. The velocity of the gantry results from backwards differencing of the gantry position. The resulting signal is then low pass filtered using a second-order digital filter.

Experimental Results The objective of the experiment is to damp out the vibration in a string subject to an impulse disturbance. First, the system is run in *open loop* with the DC motor voltage set to zero ($k_d = 0$). The open loop string angular deflection is shown in Fig. 2(a). Boundary damping with $k_d = 4.25$ achieved the best regulation results. Figure 2(b) and (c) show the boundary string angle $w_x(L,t)$ and displacement $w(L,t)$, respectively, decaying to zero more quickly than in open loop using the control voltage shown in Fig. 2(d).

Boundary Damped Cantilevered Beam

We now consider the fourth order cantilevered beam model consisting of field Eq. (2.47) and boundary conditions (2.48). If we neglect the boundary inertia ($m = 0$) and substitute boundary damping for the control force ($f = -k_d w_t(L,t)$) into the remaining boundary condition (2.49), we obtain

$$EIw_{xxx}(L,t) = k_d w_t(L,t). \tag{4.33}$$

Differentiation of the energy

$$E = \frac{1}{2}\int_0^L \left(\rho w_t^2 + EIw_{xx}^2\right) dx \geq 0 \tag{4.34}$$

produces

$$\begin{aligned}
\dot{E} &= \int_0^L \rho w_t w_{tt} + EIw_{xx}w_{xxt}dx \\
&= \int_0^L w_t\left(-EIw_{xxxx}\right) + EIw_{xx}w_{xxt}dx \\
&= -EIw_t w_{xxx}\big|_0^L + EIw_{xt}w_{xx}\big|_0^L + \int_0^L -EIw_{xxt}w_{xx} + EIw_{xx}w_{xxt}dx \\
&= -k_d w_t^2(L,t) \leq 0,
\end{aligned} \tag{4.35}$$

where integration by parts and the boundary conditions have been used. Eqs. (4.34) and (4.35) show that the boundary damped cantilevered beam is stable.

To show asymptotic stability, we can use Barbalat's Lemma, LaSalle's Invariance Principle, or develop a new Lyapunov functional that provides a negative definite derivative. Based on our experience with the boundary damped string, the Lyapunov approach has the potential of proving strong exponential stability. The trick is to find an appropriate crossing term. Let's investigate the crossing term for the boundary damped string

$$C(t) = \beta\int_0^L x w_x w_t dx, \tag{4.36}$$

where $\beta > 0$. We can bound the crossing term using Ineqs. (3.10) and (3.24) as follows:

$$\begin{aligned}
|C(t)| &= \beta\left|\int_0^L x w_x w_t dx\right| < \beta\int_0^L x\left(w_x^2 + w_t^2\right) dx \\
&\leq \beta L\int_0^L \left(w_x^2 + w_t^2\right) dx \leq \beta L\int_0^L \left(L^2 w_{xx}^2 + w_t^2\right) dx \\
&\leq \beta L\max(1,L^2)\int_0^L \left(w_{xx}^2 + w_t^2\right) dx \leq 2\beta L\frac{\max(1,L^2)}{\min(\rho,EI)}E
\end{aligned} \tag{4.37}$$

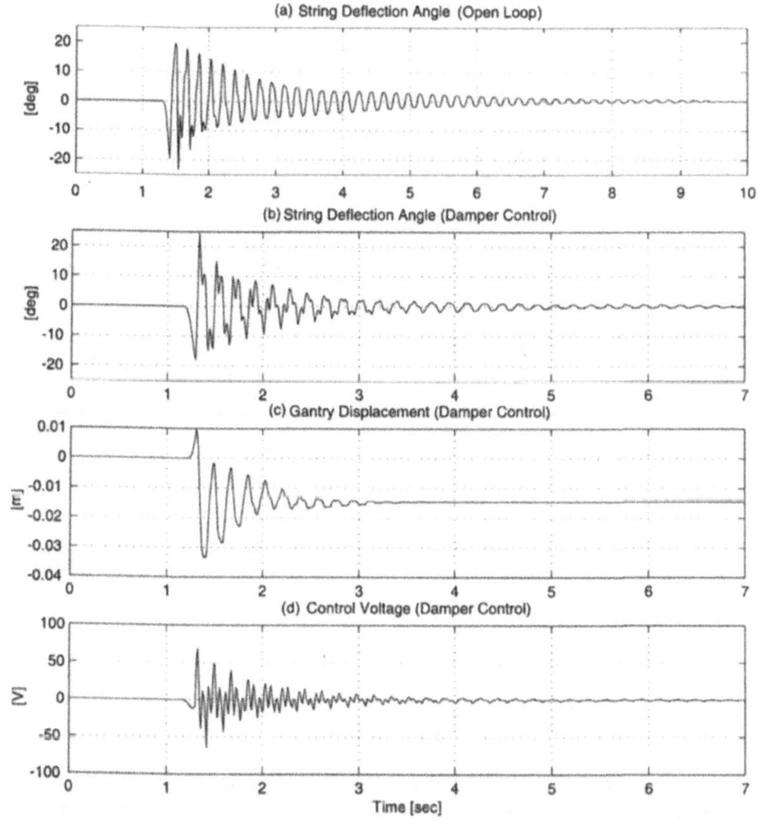

Figure 2 Experimental response of a vibrating string: (a) Open loop angle, $w_x(L,t)$; (b) Damper controlled angle, $w_x(L,t)$; (c) Damper controlled gantry displacement, $w(L,t)$; (d) Control voltage.

because

$$\int_0^L \left(w_{xx}^2 + w_t^2\right) dx \le \frac{2E}{\min(\rho, EI)}.$$

The Lyapunov functional $V = E + C$ is bounded as follows

$$\lambda_1 E(t) \le V(t) \le \lambda_2 E(t), \tag{4.38}$$

where

$$\begin{aligned} \lambda_1 &= 1 - 2\beta L \frac{\max(1, L^2)}{\min(\rho, EI)} > 0, \\ \lambda_2 &= 1 + 2\beta L \frac{\max(1, L^2)}{\min(\rho, EI)} > 1, \end{aligned}$$

provided

$$0 < \beta < \frac{\min(\rho, EI)}{2L \max(1, L^2)}. \tag{4.39}$$

The Lyapunov functional time derivative

$$\dot{V} = -k_d w_t(L,t)^2 + \dot{C} \tag{4.40}$$

involves the energy time derivative in Eq. (4.35) and

$$\dot{C} = \dot{C}_1 + \dot{C}_2, \tag{4.41}$$

where

$$\dot{C}_1 = \beta \int_0^L x w_x w_{tt} dx, \tag{4.42}$$

$$\dot{C}_2 = \beta \int_0^L x w_{xt} w_t dx = \frac{\beta}{2}\left(L w_t^2(L,t) - \int_0^L w_t^2 dx\right). \tag{4.43}$$

The term $\dot{C}_2$ simplifies as shown in Eq. (4.43) via the same steps used for the boundary controlled string. Substitution of the field Eq. (2.47) and boundary conditions (2.48) and integration by parts simplify Eq. (4.25) as follows

$$\begin{aligned} \dot{C}_1 &= \beta \int_0^L x w_x \left(-\frac{EI}{\rho} w_{xxxx}\right) dx \\ &= \frac{\beta EI}{\rho}\left(- x w_x w_{xxx}\Big|_0^L + \int_0^L \left(w_x + x w_{xx}\right) w_{xxx} dx\right) \qquad (4.44) \\ &= \frac{\beta EI}{\rho}\left(-L w_x(L) w_{xxx}(L) + w_x w_{xx} + x w_{xx}^2\Big|_0^L \right. \\ &\quad \left. - \int_0^L 2 w_{xx}^2 + x w_{xxx} w_{xx} dx\right). \qquad (4.45) \end{aligned}$$

All of the boundary terms cancel in Eqs. (4.44) and (4.45) except $-Lw_x(L)w_{xxx}(L)$. Thus, we may add Eqs. (4.44) and (4.45) to form

$$\begin{aligned} 2\dot{C}_1 &= \frac{\beta EI}{\rho}\left(-2Lw_x(L)w_{xxx}(L) - \int_0^L 2w_{xx}^2 - w_x w_{xxx} dx\right) \\ &= \frac{\beta EI}{\rho}\left(-2Lw_x(L)w_{xxx}(L) + w_x w_{xx}\big|_0^L - \int_0^L 3w_{xx}^2 dx\right). \end{aligned} \tag{4.46}$$

Now we combine Eqs. (4.35), (4.46), and (4.43) to produce

$$\begin{aligned} \dot{V} &= -k_d w_t^2(L) + \frac{\beta}{2}\left(Lw_t^2(L) - \int_0^L w_t^2 dx\right) \\ &\quad + \frac{\beta EI}{2\rho}\left(-2Lw_x(L)w_{xxx}(L) - \int_0^L 3w_{xx}^2 dx\right) \\ &= -\left(k_d - \frac{\beta L}{2}\right) w_t^2(L) - \frac{\beta L k_d}{\rho} w_x(L) w_t(L) \\ &\quad - \frac{\beta EI}{\rho}\int_0^L w_{xx}^2 dx - \frac{\beta}{2}\int_0^L w_t^2 + \frac{EI}{\rho} w_{xx}^2 dx \end{aligned} \tag{4.47}$$

where we have used boundary condition (4.33). Substitution of the energy Eq. (4.34) and application of the nonlinear damping inequality (3.12) and the integral downshift inequality (3.9) to Eq. (4.47) produce

$$\begin{aligned} \dot{V} &\le -\left(k_d - \frac{\beta L}{2}\right) w_t^2(L) + \frac{\beta L k_d}{\rho}\left(\delta w_x^2(L) + \frac{1}{\delta} w_t^2(L)\right) - \frac{\beta EI}{\rho L} w_x^2(L) - \frac{\beta}{\rho} E \\ &= -\left(k_d - \frac{\beta L}{2} - \frac{\beta L k_d}{\rho \delta}\right) w_t^2(L) - \frac{\beta}{\rho}\left(\frac{EI}{L} - \delta\right) w_x^2(L) - \frac{\beta}{\rho} E \\ &\le -\frac{\beta}{\rho \lambda_2} V \end{aligned} \tag{4.48}$$

provided we choose

$$\begin{aligned} \delta &< \frac{EI}{\rho}, \\ \beta &< \frac{2\rho \delta k_d}{\rho \delta L + 2L k_d}. \end{aligned}$$

The bounding solution of Ineq. (4.48) is

$$V(t) \le V(0) e^{-\frac{\beta}{\rho \lambda_2} t}. \tag{4.49}$$

Use of Ineq. (4.38) in Ineq. (4.49) produces

$$\int_0^L w_{xx}^2 dx \le \frac{2}{EI} E \le \frac{2}{EI\lambda_1} V \le \frac{2V(0)}{EI\lambda_1} e^{-\frac{\beta}{\rho\lambda_2} t}. \tag{4.50}$$

Ineqs. (3.24) and (3.5) relate the pointwise displacement to the integral in Ineq. (4.50)

$$\int_0^L w_{xx}^2 dx \geq \frac{1}{L^2}\int_0^L w_x^2 dx \geq \frac{1}{L^3}w^2(x,t).$$

Thus, the following theorem follows directly.

Theorem 26 *The boundary damped cantilevered beam described by Eqs. (2.47), (2.48), and (4.33) is strongly exponential stable.*

4.1.2 Distributed Damping

Environmental and material damping exists in most distributed systems. Environmental damping results from motion of the material through a dissipative fluid such as air or water. Often a linear viscous model for external damping with a distributed force proportional to material velocity can accurately account for the energy dissipation associated with fluid viscosity and/or aerodynamic drag effects. Material damping can be modeled as distributed forces proportional to material strain rates. More accurate models involve nonlinearity and/or the viscoelastic behavior of the material and can add significant complexity to model-based control development. Accurate quantification of damping is difficult because of the viscous damping model approximations, measurement complexity, and variability of the damping constants with material age and environmental conditions. It is therefore advisable to neglect damping during control development to obtain a conservative stability result. In many cases, however, inclusion of damping in the system model is required to obtain a stability result and/or greatly simplify the control proof. In this section, we show that distributed viscous damping ensures bounded response to distributed forcing and boundary disturbances.

Disturbed String with Distributed Damping

Figure 3 shows a linear string model with distributed viscous damping coefficient b, disturbance force $f(x,t)$, and prescribed boundary disturbances $w_0(t)$ and $w_L(t)$ as follows

$$\rho w_{tt} + bw_t - Pw_{xx} = f \text{ for } x \in (0,L), \tag{4.51}$$
$$w(0,t) = w_0(t), \tag{4.52}$$
$$w(L,t) = w_L(t). \tag{4.53}$$

We introduce the transformation

$$w^*(x,t) = w(x,t) - \frac{x}{L}w_L(t) + \frac{x-L}{L}w_0(t)$$

to obtain pinned boundary conditions

$$w^*(0,t) = w^*(L,t) = 0 \tag{4.54}$$

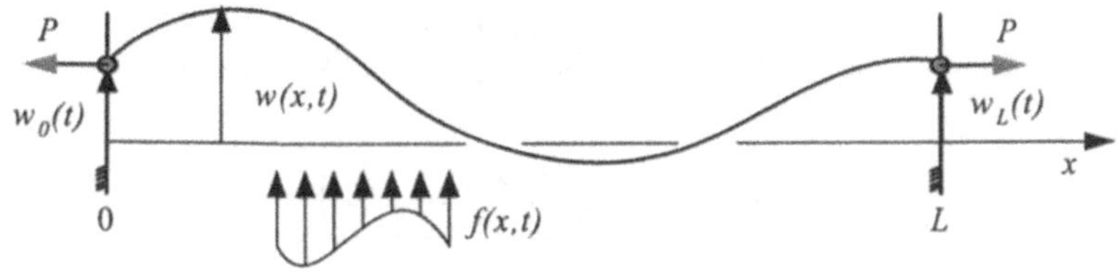

Figure 3 Schematic diagram of the disturbed string with distributed damping.

and modify the field equation to

$$\rho w_{tt}^* + b w_t^* - P w_{xx}^* = f^* \text{ for } x \in (0, L), \tag{4.55}$$

where

$$f^*(x,t) = f(x,t) - \frac{x}{L}\left(\rho \ddot{w}_L(t) + b\dot{w}_L(t)\right) + \frac{x-L}{L}\left(\rho \ddot{w}_0(t) + b\dot{w}_0(t)\right). \tag{4.56}$$

We assume that $f^*(x,t) \in \mathcal{L}_\infty$ so the distributed force, boundary velocities, and boundary accelerations must be bounded. In addition, we assume that the boundary displacements are bounded so that $w^*(x,t) \in \mathcal{L}_\infty$ implies $w(x,t) \in \mathcal{L}_\infty$.

Theorem 27 *If $f^*(x,t) \in \mathcal{L}_\infty$ in Eq. (4.55) with boundary conditions Eq. (4.54) then $w^*(x,t) \in \mathcal{L}_\infty$ for all $x \in [0,L]$ and $t \in [0,\infty)$.*

Proof: For simplicity we drop the star notation. We use the Lyapunov functional given in Eq. (4.16) with a different crossing term

$$C = \alpha\rho \int_0^L w w_t dx, \tag{4.57}$$

where α is a sufficiently small positive constant. Application of Ineqs. (3.10) and (3.16) to Eq. (4.57) yields:

$$\begin{aligned} |C(t)| &\leq \alpha\rho \int_0^L \left(w^2 + w_t^2\right) dx \leq \alpha\rho \int_0^L \left(w_t^2 + L^2 w_x^2\right) dx \\ &\leq 2\alpha\rho \frac{\max(1, L^2)}{\min(\rho, P)} E(t) \end{aligned} \tag{4.58}$$

so

$$-2\alpha\rho \frac{\max(1,L^2)}{\min(\rho,P)} E(t) \leq C(t) \leq 2\alpha\rho \frac{\max(1,L^2)}{\min(\rho,P)} E(t). \tag{4.59}$$

Adding $E(t)$ to Ineq. (4.59) results in

$$0 \leq \lambda_1 E(t) \leq V(t) \leq \lambda_2 E(t), \tag{4.60}$$

where

$$\lambda_1 = 1 - 2\alpha\rho \frac{\max(1, L^2)}{\min(\rho, P)} > 0 \tag{4.61}$$

and

$$\lambda_2 = 1 + 2\alpha\rho \frac{\max(1, L^2)}{\min(\rho, P)} > 1 \tag{4.62}$$

if we choose

$$\alpha < \frac{\min(\rho, P)}{2\rho \max(1, L^2)}.$$

The time derivative of $V(t)$ depends on

$$\dot{E} = -\int_0^L b w_t^2 dx + \int_0^L w_t f dx, \tag{4.63}$$

where the field Eq. (4.55), boundary conditions (4.54), and integration by parts have been applied. Application of Ineq. (3.12) produces

$$\dot{E} \le -(b - \delta) \int_0^L w_t^2 dx + \frac{1}{\delta} \int_0^L f^2 dx, \tag{4.64}$$

where $\delta > 0$ is a positive constant.

The remaining term in $\dot{V}(t)$ is

$$\begin{aligned} \dot{C} &= \alpha\rho \int_0^L \left(w_t^2 + w w_{tt}\right) dx \\ &= \alpha \int_0^L \rho w_t^2 dx + \alpha \int_0^L w \left(P w_{xx} + f - b w_t\right) dx \end{aligned} \tag{4.65}$$

where Eq. (4.51) has been used. The terms in Eq. (4.65) simplify via integration by parts, the boundary conditions, and Ineqs. (3.12) and (3.16) as follows

$$\int_0^L w P w_{xx} dx = -P \int_0^L w_x^2 dx, \tag{4.66}$$

$$-b \int_0^L w w_t dx \le \frac{b}{\delta} \int_0^L w_t^2 dx + \delta b L^2 \int_0^L w_x^2 dx, \tag{4.67}$$

$$\int_0^L w f dx \le \frac{1}{\delta} \int_0^L f^2 dx + L^2 \delta \int_0^L w_x^2 dx. \tag{4.68}$$

Combination of Eqs. (4.64) – (4.68) yields

$$\begin{aligned} \dot{V}(t) \le & -\left(b - \delta - \alpha\left(\rho + \frac{b}{\delta}\right)\right) \int_0^L w_t^2 dx + \frac{1}{\delta}(\alpha + 1) \int_0^L f^2 dx \\ & -\alpha\left(P - \delta L^2 (b+1)\right) \int_0^L w_x^2 dx \le -\lambda_3 E(t) + \varepsilon, \end{aligned} \tag{4.69}$$

where

$$\lambda_3 = \frac{\min\left(b - \delta - \alpha\left(\rho + \frac{b}{\delta}\right), \alpha\left(P - \delta L^2\left(b+1\right)\right)\right)}{\max\left(P, \rho\right)} \tag{4.70}$$

and

$$\varepsilon = \frac{1}{\delta}\left(\alpha + 1\right) \max_{t=[0,\infty)} \left(\int_0^L f^2\left(x,t\right) dx\right). \tag{4.71}$$

The assumption that $f(x,t) \in \mathcal{L}_\infty$ implies that $\varepsilon < \infty$. Use of Ineq. (4.60) gives

$$\dot{V} \leq -\lambda V + \varepsilon, \tag{4.72}$$

where

$$\lambda = \frac{\lambda_3}{\lambda_2} > 0 \tag{4.73}$$

if α and δ are sufficiently small. Solution of Eq. (4.72) produces

$$V\left(t\right) \leq V\left(0\right) e^{-\lambda t} + \frac{\varepsilon}{\lambda} \tag{4.74}$$

so the Lyapunov functional is bounded for all $t \in [0, \infty)$.

Using Eqs. (4.60) and (4.5) and Ineq. (3.8),

$$\frac{\lambda_1 P}{2L} w^2(x,t) \leq \frac{\lambda_1 P}{2} \int_0^L w_x^2 dx \leq \lambda_1 E \leq V(t) \leq V\left(0\right) e^{-\lambda t} + \frac{\varepsilon}{\lambda} \in \mathcal{L}_\infty \tag{4.75}$$

Thus $w(x,t) \in \mathcal{L}_\infty$ for $x \in [0, L]$ and $t \in [0, \infty)$.
□

Disturbed Beam with Distributed Damping

Consider the linear beam model shown in Fig. 4 with distributed viscous damping coefficient b, disturbance force $f(x,t)$, and prescribed boundary displacement ($w_0\left(t\right)$ and $w_L\left(t\right)$) and rotation ($w_{x0}(t)$ and $w_{xL}(t)$) disturbances as follows

$$\rho w_{tt} + b w_t + EI w_{xxxx} = f \text{ for } x \in (0, L), \tag{4.76}$$

$$w(0,t) = w_0\left(t\right), \tag{4.77}$$

$$w_x(0,t) = w_{x0}(t), \tag{4.78}$$

$$w(L,t) = w_L\left(t\right), \tag{4.79}$$

$$w_x(L,t) = w_{xL}(t). \tag{4.80}$$

We introduce the transformation

$$w^*(x,t) = w(x,t) - (a_0 + a_1 x + a_2 x^2 + a_3 x^3),$$

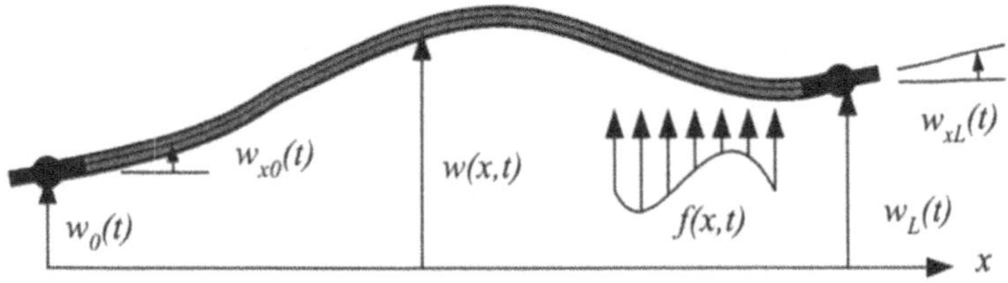

Figure 4 Schematic diagram of the disturbed beam with distributed damping.

where the coefficients $a_0, \ldots, a_3$ are chosen to ensure clamped boundary conditions on $w^*(x,t)$:

$$\begin{aligned} w^*(0,t) &= w_0 - a_0 = 0, && (4.81)\\ w_x^*(0,t) &= w_{x0} - a_1 = 0, && (4.82)\\ w^*(L,t) &= w_L - (a_0 + a_1 L + a_2 L^2 + a_3 L^3) = 0, && (4.83)\\ w_x^*(L,t) &= w_{xL} - (a_1 + 2a_2 L + 3a_3 L^2) = 0. && (4.84)\end{aligned}$$

The coefficients that satisfy Eqs. (4.81) - (4.84) are $a_0 = w_0$, $a_1 = w_{x0}$, $a_2 = -\frac{1}{L^2}(Lw_{xL} + 2Lw_{x0} - 3w_L + 3w_0)$, $a_3 = \frac{1}{L^3}(-2w_L + 2w_0 + Lw_{x0} + Lw_{xL})$. The transformed field equation is

$$\rho w_{tt}^* + b w_t^* + EI w_{xxxx}^* = f^* \text{ for } x \in (0, L), \tag{4.85}$$

where

$$f^*(x,t) = f(x,t) - \rho(\ddot{a}_0 + \ddot{a}_1 x + \ddot{a}_2 x^2 + \ddot{a}_3 x^3) - b(\dot{a}_0 + \dot{a}_1 x + \dot{a}_2 x^2 + \dot{a}_3 x^3). \tag{4.86}$$

If we assume that $f^*(x,t) \in \mathcal{L}_\infty([0,L])$ then the distributed force and boundary translational and rotational velocities and accelerations must be bounded. In addition, if we assume that the boundary displacements and rotations are bounded then $w^*(x,t) \in \mathcal{L}_\infty([0,L])$ implies $w(x,t) \in \mathcal{L}_\infty([0,L])$.

Theorem 28 *The disturbed beam governed by field Eq. (4.76) and boundary conditions (4.77) - (4.80) is strongly stable if* $b \neq 0$ *and* $\ddot{w}$, $\dot{w}$, w, $\ddot{w}_x$, $\dot{w}_x$, *and* $w_x \in \mathcal{L}_\infty$ *at* $x = 0$ *and* L *and* $f \in \mathcal{L}_\infty([0,L])$.

Proof: With the theorem assumptions, $w^*(x,t) \in \mathcal{L}_\infty([0,L])$ means strong stability for $w(x,t)$ so we focus on the transformed field Eq. (4.85) and boundary conditions (4.81) - (4.84). For simplicity, we drop the *'s in the proof. A Lyapunov functional is chosen to be

$$V(t) = E_b(t) + E_c(t), \tag{4.87}$$

where

$$E_b = \frac{1}{2}\rho \int_0^L w_t^2 dx + \frac{1}{2} EI \int_0^L w_{xx}^2 dx, \tag{4.88}$$

and the crossing term

$$E_c = \beta\rho \int_0^L w w_t dx, \tag{4.89}$$

with $\beta > 0$ a small positive constant. Since $w(0) = 0$, we can use Ineq. (3.23) in Eq. (4.88) to produce

$$E_b \geq \frac{1}{2}\rho \int_0^L w_t^2 dx + \frac{\pi^2 EI}{8L^2} \int_0^L w_x^2 dx. \tag{4.90}$$

Substitution of Ineqs. (3.23) and (3.10) into Eq. (4.89) yields:

$$\begin{aligned} |E_c(t)| &\leq \beta\rho \int_0^L \left(w_t^2 + w^2\right) dx \leq \beta\rho \int_0^L \left(w_t^2 + \frac{4L^2}{\pi^2} w_x^2\right) dx \\ &\leq 2\beta\rho \frac{\max\left(1, \frac{4L^2}{\pi^2}\right)}{\min\left(\rho, \frac{\pi^2}{4L^2} EI\right)} E_b(t). \end{aligned} \tag{4.91}$$

If we choose

$$\beta \leq \frac{\min\left(\rho, \frac{\pi^2}{4L^2} EI\right)}{2\rho \max\left(1, \frac{4L^2}{\pi^2}\right)}$$

then

$$0 \leq \lambda_1 E_b(t) \leq V(t) \leq \lambda_2 E_b(t), \tag{4.92}$$

where

$$\lambda_1 = 1 - 2\beta\rho \frac{\max\left(1, \frac{4L^2}{\pi^2}\right)}{\min\left(\rho, \frac{\pi^2}{4L^2} EI\right)} > 0, \tag{4.93}$$

$$\lambda_2 = 1 + 2\beta\rho \frac{\max\left(1, \frac{4L^2}{\pi^2}\right)}{\min\left(\rho, \frac{\pi^2}{4L^2} EI\right)} > 1. \tag{4.94}$$

The time derivative of $V(t)$

$$\dot{V}(t) = \dot{E}_b(t) + \dot{E}_c(t) \tag{4.95}$$

includes the net energy rate of change

$$\begin{aligned} \dot{E}_b &= \int_0^L \rho w_t w_{tt} + EI w_{xx} w_{xxt} dx \\ &= EI\left(w_{xx} w_{xt} - w_{xxx} w_t\right)\Big|_0^L + \int_0^L w_t\left(-b w_t - EI w_{xxxx} + f\right) + EI w_{xxxx} w_t dx \\ &= -\int_0^L b w_t^2 dx + \int_0^L w_t f dx, \end{aligned} \tag{4.96}$$

where the field Eq. (4.85), boundary conditions (4.81) - (4.82), and integration by parts have been applied. Application of Ineq. (3.12) produces

$$\dot{E}_b \leq -(b-\delta_1)\int_0^L w_t^2 dx + \frac{1}{\delta_1}\int_0^L f^2 dx, \tag{4.97}$$

where δ_1 is an arbitrary positive constant.

The cross term power is

$$\begin{aligned} \dot{E}_c &= \beta\rho\int_0^L \left(w_t^2 + ww_{tt}\right) dx \\ &= \beta\int_0^L -EIww_{xxxx} - bww_t + wf + \rho w_t^2 dx. \end{aligned} \tag{4.98}$$

The first term in Eq. (4.98) simplifies via integration by parts and the boundary conditions to

$$\begin{aligned} -\beta EI\int_0^L ww_{xxxx}dx &= \beta EI\left(ww_{xxx} - w_x w_{xx}\right)\Big|_0^L - \beta EI\int_0^L w_{xx}^2 dx \\ &= -\beta EI\int_0^L w_{xx}^2 dx. \end{aligned} \tag{4.99}$$

Using Ineqs. (3.23) and (3.12), the second and third terms in Eq. (4.98) become

$$\begin{aligned} -\beta b\int_0^L ww_t dx &\leq \beta\frac{b}{\delta_2}\int_0^L w_t^2 dx + \beta b\delta_2\frac{4L^2}{\pi^2}\int_0^L w_x^2 dx \\ &\leq \beta\frac{b}{\delta_2}\int_0^L w_t^2 dx + \beta b\delta_2\frac{16L^4}{\pi^4}\int_0^L w_{xx}^2 dx, \end{aligned} \tag{4.100}$$

$$\begin{aligned} \beta\int_0^L wf dx &\leq \frac{\beta}{\delta_3}\int_0^L f^2 dx + \beta\frac{4L^2}{\pi^2}\delta_3\int_0^L w_x^2 dx \\ &\leq \frac{\beta}{\delta_3}\int_0^L f^2 dx + \beta\frac{16L^4}{\pi^4}\delta_3\int_0^L w_{xx}^2 dx, \end{aligned} \tag{4.101}$$

where δ_2 and δ_3 are arbitrary positive constants.

Substitution of Eqs. (4.99) through (4.101) into Eq. (4.95) yields:

$$\begin{aligned} \dot{V}(t) \leq & -\left(b-\delta_1-\beta\rho-\frac{\beta b}{\delta_2}\right)\int_0^L w_t^2 dx \\ & -\beta\left(EI - b\delta_2\frac{16L^4}{\pi^4} - \frac{16L^4}{\pi^4}\delta_3\right)\int_0^L w_{xx}^2 dx \\ & +\left(\frac{1}{\delta_1}+\frac{\beta}{\delta_3}\right)\int_0^L f^2 dx \\ \leq & -\lambda_3 E_b(t) + \varepsilon, \end{aligned} \tag{4.102}$$

where

$$\lambda_3 = \min\left(\frac{\varepsilon_1}{\rho}, \frac{\varepsilon_2}{EI}, \beta\right) > 0, \tag{4.103}$$

$$\varepsilon = \left(\frac{1}{\delta_1} + \frac{\beta}{\delta_3}\right) \max_{t=[0,\infty)} \left(\int_0^L f^2 dx\right) < \infty, \tag{4.104}$$

$$\varepsilon_1 = b - \delta_1 - \beta\rho - \frac{\beta b}{\delta_2} > 0, \tag{4.105}$$

$$\varepsilon_2 = \beta\left(EI - b\delta_2\frac{16L^4}{\pi^4} - \frac{16L^4}{\pi^4}\delta_3\right) > 0, \tag{4.106}$$

if β, $\delta_1, \delta_2,$ and δ_3 are sufficiently small.

Use of Eq. (4.92) produces

$$\dot{V}(t) \leq -\lambda V(t) + \varepsilon, \tag{4.107}$$

where $\lambda = \lambda_3/\lambda_2$, with the solution

$$V(t) \leq V(0)\, e^{-\lambda t} + \frac{\varepsilon}{\lambda} \in \mathcal{L}_\infty.$$

Using Ineq. (3.5) in Eq. (4.92) then yields,

$$\frac{\pi^2 EI}{8L^3}\lambda_1 w^2 \leq \frac{\pi^2 EI}{8L^2}\lambda_1 \int_0^L w_x^2 dx \leq \lambda_1 \frac{EI}{2} \int_0^L w_{xx}^2 dx \leq \lambda_1 E_b \leq V(t) \in \mathcal{L}_\infty. \tag{4.108}$$

so $w \in \mathcal{L}_\infty([0, L])$.

□

4.1.3 Pointwise Damping

Domain Controlled String

In Chapter 3 we showed that a pinned string with pointwise damping feedback modeled by Eqs. (3.27) - (3.28) generates a strongly continuous C_0 semigroup. We now apply LaSalle's Invariance Principle to show that if the point damper is located with x_c an irrational number, then asymptotic stability is guaranteed. First, we show precompactness of solution trajectories.

Lemma 29 *For the operator $\mathcal{A}$ in Eq. (3.32) $0 \in \mathcal{R}(\mathcal{A})$ and $(\mathcal{I} - \mathcal{A})^{-1}$ is compact.*

Proof: Clearly, $0 \in \mathcal{R}(A)$. For the compactness of $(\mathcal{I} - \mathcal{A})^{-1}$, let $\{\mathbf{Y}_n\} \subset \mathcal{H}$ be a bounded sequence and denote $\mathbf{z}_n = (\mathcal{I} - \mathcal{A})^{-1}\mathbf{Y}_n$ for $n \geq 1$. Due to the dissipativity of $\mathcal{A}$, there exists a constant $c_1 > 0$ such that

$$\|\mathbf{z}_n\| \leq \|\mathbf{Y}_n\| \leq c_1, \qquad \|A\mathbf{z}_n\| \leq \|\mathbf{Y}_n\| \leq c_1, \qquad \forall n \geq 1.$$

Setting $\mathbf{z}_n = [w_n, \nu_n]^T$, then there exists a constant $c_2 > 0$ such that

$$\|w_n\|_{H^2} \le c_2, \quad \|\nu_n\|_{H^1} \le c_2, \quad \forall n \ge 1,$$

where $\|\cdot\|_{H^k}$ is the norm in the space $H^k(0, x_c^-) \cup (x_c^+, L)$. By the Sobolev embedding theorem (see [115] for details), there exists a subsequence of $\{\mathbf{z}_n\}$, denoted again by $\{\mathbf{z}_n\}$, such that $w_n \to w$ in $H^1(0, x_c^-) \cup (x_c^+, L), \nu_n \to \nu$ in $C^0((0, x_c^-) \cup (x_c^+, L))$. Therefore, we get $\mathbf{z}_n \to \mathbf{z}$ as $n \to \infty$, where $\mathbf{z} = [w, \nu]^T$.
□

Theorem 30 *The energy $E(t)$ of the closed loop system given in Eq. (3.29), for any initial data $\mathbf{z}_0 \in \mathcal{H}$, decays to zero asymptotically with time.*

Proof: To prove the theorem, we use LaSalle's invariance principle (Theorem 24). According to this principle, all solutions of the system asymptotically tend to the maximal invariant subset of the following set:

$$\mathcal{S} = \left\{ \mathbf{z} \in \mathcal{H} | \dot{E} = 0 \right\}$$

provided that the closure of the solution trajectories for $t \ge 0$ are compact in $\mathcal{H}$. Thus by Lemma 29, it is enough to prove that, for an initial data $\mathbf{z}_0 = [w_0, \nu_0]^T \in \mathcal{D}(\mathcal{A})$, the condition $\dot{E}(t) = 0, \forall\, t \ge 0$ implies $\mathbf{z}_0 = [0, 0]^T$. The result can be easily extended to any $\mathbf{z}_0 \in \mathcal{H}$ because $\mathcal{A}$ is densely defined and generates a semigroup solution. From $\dot{E}(t) = 0, \forall t \ge 0$, we know that $w_t(x_c, t) = 0, \forall\, t \ge 0$ so $\mathbf{z}(t) = [w, \nu]^T$ satisfies

$$\begin{aligned} &\rho w_{tt} - P w_{xx} = 0, \\ &w_t(x_c) = 0, \\ &w(0) = 0, \\ &w(L) = 0. \end{aligned} \tag{4.109}$$

The boundary conditions are separable so the solution of Eqs. (4.109) can be found using separation of variables [81]. We assume the solution has the following form:

$$w(x, t) = W(x)\eta(t). \tag{4.110}$$

The condition $w_t(x_c, t) = 0$ results from either

i) $\dot{\eta}(t) \equiv 0$. From the field equation, it follows that $W(x) = c_1 x + c_2$, where c_1 and c_2 are arbitrary constants. Substitution into Eq. (4.109), we obtain $c_1 = c_2 = 0$ and $w(x, t) = 0$.

ii) $\dot{\eta}(t) \neq 0$. Substitution of (4.110) into (4.109), we obtain the following solution for $W(x)$:

$$W(x) = c_1 \cos(\beta x) + c_2 \sin(\beta x)$$

where c_1, c_2 and β are constants to be determined. From $w(0) = 0$ we obtain $c_1 = 0$. From $w(L) = 0$ we have $c_2 = 0$ or $\beta = \frac{k\pi}{L}, k = 0, 1, \ldots, \infty$. If $c_2 = 0$, we have

$W(x) = 0$. If $c_2 \neq 0$, $w_t(x_c) = 0$ requires that $\beta = 0$. So we obtain the result $w(x,t) = 0$.
□

Using inequalities we can show that the weak result of Theorem 30 implies strong asymptotic stability for the string displacement.

4.1.4 Parametric Damping

An axial damper can asymptotically stabilize the transverse vibration of the cantilevered beam model described by field Eq. (2.84) and boundary conditions (2.81). The parametric damping control law is

$$F_1(t) = -k_d \dot{y}(t) = \frac{k_d}{2} \int_0^1 \left[w_x^2\right]_t dx. \tag{4.111}$$

where k_d is a positive control gain and Eq. (2.85) has been used. The energy given by

$$E = \frac{1}{2} \int_0^1 w_t^2 + F_0 w_x^2 + w_{xx}^2 dx \geq 0 \tag{4.112}$$

has a time derivative

$$\begin{aligned} \dot{E} &= \int_0^1 w_t \left(F_0 w_{xx} - w_{xxxx} + F_1(t) w_{xx}\right) + F_0 w_x w_{xt} + w_{xx} w_{xxt} dx \\ &= \left(F_0 w_x w_t - w_{xxx} w_t + w_{xx} w_{xt} + F_1 w_x w_t\right)\Big|_0^1 - F_1 \int_0^1 w_{xt} w_x dx \\ &= -k_d \dot{y}^2 \leq 0, \end{aligned} \tag{4.113}$$

where the field Eq. (2.84), boundary conditions (2.81), and control law (4.111) have been used. Clearly, the closed loop system is stable ($E \in \mathcal{L}_\infty$). Using Ineq. (3.5) and (3.9), we can show that $w(x,t), w_x(x,t) \in \mathcal{L}_\infty(0,1)$. If we assume $w \in \mathcal{S}_\mathcal{A}(0,1)$, then boundedness of the displacement implies $w_{xx}(x,t), w_{xxxx}(x,t) \in \mathcal{L}_\infty(0,1)$.

We will use two approaches to prove asymptotic stability. First, we apply Barbalat's Lemma. The second derivative of the energy is

$$\ddot{E} = -k_d \dot{y} \ddot{y}. \tag{4.114}$$

We can bound the axial velocity

$$\begin{aligned} |\dot{y}| &= \left| \int_0^1 w_{xt} w_x dx \right| = \left| \int_0^1 w_t w_{xx} dx \right| \\ &\leq \int_0^1 w_t^2 + w_{xx}^2 dx \in \mathcal{L}_\infty. \end{aligned} \tag{4.115}$$

using integration by parts, the boundary conditions, and Ineq. (3.10). The field Eq. (2.84) and $F_1 \in \mathcal{L}_\infty$ then imply $w_{tt}(x,t) \in \mathcal{L}_\infty(D)$. The axial acceleration

$$\begin{aligned} |\ddot{y}| &= \left| \int_0^1 [w_{xt} w_x]_t \, dx \right| = \left| \int_0^1 w_{xtt} w_x + w_{xt}^2 dx \right| \\ &= \left| \int_0^1 w_{tt} w_{xx} dx \right| + \left| \int_0^1 2 w_t w_{xxt} dx \right| \end{aligned} \quad (4.116)$$

is bounded if we assume $w_t \in \mathcal{S}_{\mathcal{A}}(0,L) \bigcap \mathcal{S}(0,L)$ which implies that $w_t(x,t), w_{xxt}(x,t) \in \mathcal{L}_\infty(0,1)$. Thus, the energy decays to zero and

$$\int_0^1 \left[w_x^2\right]_t dx \to 0 \text{ as } t \to \infty. \quad (4.117)$$

Unfortunately, there are no inequalities from Chapter 3 that allow us to obtain a weak asymptotic stability result from Eq. (4.117).

The nonlinear closed loop system described by Eqs. (2.84) and (4.111) produces a semigroup solution so we may apply LaSalle's Invariance Principle [6]. If we substitute the separable solution

$$w(x,t) = \sum_{k=1}^{\infty} W_k(x) \eta_k(t)$$

into Eq. (4.113), we obtain

$$\int_0^1 \left[\left(\sum_{k=1}^{\infty} W_{kx} \eta_k \right)^2 \right]_t dx = 2 \int_0^1 \left(\sum_{k=1}^{\infty} W_{kx} \eta_k \right) \left(\sum_{l=1}^{\infty} W_{lx} \dot{\eta}_l \right) dx \equiv 0. \quad (4.118)$$

If Eq. (4.118) implies that $\eta_k \equiv 0$ for all k, then the closed loop system is asymptotically stable. For the clamped beam, the mode functions $W_k(x)$ are not orthogonal with respect to the tension operator and cross terms in Eq. (4.118) make analysis difficult. For the pinned beam, however, we have the mode functions

$$W_k(x) = \sin(k\pi x)$$

with the orthogonality condition

$$\int_0^1 W_{kx} W_{lx} dx = \frac{k^2 \pi^2}{2} \delta_{kl}$$

so the cross terms cancel and Eq. (4.118) simplifies to

$$\sum_{k=1}^{\infty} W_{kx} \eta_k \dot{\eta}_k \equiv 0. \quad (4.119)$$

The eigenvalues of a pinned beam are distinct so the $\eta_k(t)$ are sinusoidal at different frequencies and therefore linearly independent. The only way Eq. (4.119) can be satisfied for all t is for all $\eta_k(t) \equiv 0$. Thus, we have proven asymptotic stability for $w(x,t)$.

4.2 Passive Boundary Control

4.2.1 Boundary Controlled String

Actuator dynamics in the form of boundary inertia complicate the stability analysis. In this section, we develop passive dynamic compensators for the pinned-controlled string model described by Eqs. (2.14) - (2.16) including the effects of actuator mass. The passive boundary control law is defined to be

$$F(s) = -H_1(s)\Theta(s) - H_2(s)V(s), \tag{4.120}$$

where $F(s)$, $\Theta(s)$, and $V(s)$ are the Laplace Transforms of $f(t)$, $w_x(L,t)$, and $w_t(L,t)$, respectively. Thus, the control law depends on measurement of the boundary velocity and slope or angle. Substitution of Eq. (4.120) into the Laplace Transform of the natural boundary condition (2.16) yields the transfer function

$$G(s) = -\frac{V(s)}{\Theta(s)} = \frac{H_1(s) + P}{H_2(s) + ms}. \tag{4.121}$$

The control transfer functions $H_1(s)$ and $H_2(s)$ are chosen so that $G(s)$ is strictly stable and satisfies

$$\mathcal{R}\{G(j\omega)\} > \gamma > 0 \qquad \forall\ \omega \geq 0. \tag{4.122}$$

This condition is slightly more restrictive than the well-known strictly positive real condition $\mathcal{R}\{G(j\omega)\} > 0$ [118].

Control Development The Lyapunov analysis begins with the familiar functionals

$$E = \frac{1}{2}\int_0^L \left(\rho w_t^2 + P w_x^2\right) dx, \tag{4.123}$$

$$C = \beta \int_0^L x w_t w_x dx, \tag{4.124}$$

representing the string energy (not including boundary mass kinetic energy) and a crossing term, respectively. Differentiation with respect to time of Eq. (4.123), substitution of Eq. (2.14), and integration by parts yield

$$\dot{E} = P w_t(L,t) w_x(L,t). \tag{4.125}$$

Differentiation of Eq. (4.124) with respect to time yields

$$\dot{C} = \beta \int_0^L x w_x w_{tt} dx + \beta \int_0^L x w_{xt} w_t dx. \tag{4.126}$$

Substitution of the field Eq. (2.14) and integration by parts for the two terms in Eq. (4.126) yield

$$\mathcal{I}_1 = \beta\int_0^L x w_x w_{tt} dx = \frac{\beta P}{\rho}\left[L w_x^2(L,t) - \int_0^L w_x^2 dx\right] - \mathcal{I}_1, \tag{4.127}$$

$$\mathcal{I}_2 = \beta\int_0^L x w_{xt} w_t dx = \beta\left[L w_t^2(L,t) - \int_0^L w_t^2 dx\right] - \mathcal{I}_2. \tag{4.128}$$

Solution of Eqs. (4.127) and (4.128) and substitution into Eq. (4.126) yield

$$\dot{C} = \frac{\beta}{2}\left(L w_t^2(L,t) + \frac{PL}{\rho} w_x^2(L,t) - \int_0^L \left(w_t^2 + \frac{P}{\rho} w_x^2\right) dx\right). \tag{4.129}$$

Theorem 31 *Given Eq. (2.14) with boundary conditions (2.15) and (2.16), the boundary controller (4.120) ensures* $w(x,t) \to 0$ *exponentially fast, if* $G(s)$ *in Eq. (4.121) satisfies Eq. (4.122).*

Proof: Consider the positive definite function:

$$V_z = P\,\mathbf{x}^T\mathbf{P}\mathbf{x}, \tag{4.130}$$

where $\mathbf{P} = \mathbf{P}^T > 0$. The state space realization of $G(s)$

$$\dot{\mathbf{x}} = \mathbf{A}\mathbf{x} + \mathbf{b}w_x(L,t), \tag{4.131}$$

$$y = \mathbf{c}^T\mathbf{x} + d w_x(L,t), \tag{4.132}$$

includes the state $\mathbf{x}(t)$ and the constant matrices $\mathbf{A}$, $\mathbf{b}$, $\mathbf{c}$, and d where

$$G(s) = \mathbf{c}^T(s\mathbf{I} - \mathbf{A})^{-1}\mathbf{b} + d. \tag{4.133}$$

Differentiation with respect to time of Eq. (4.130) and substitution of Eq. (4.131) yield

$$\dot{V}_z = P\left(\mathbf{x}^T\left(\mathbf{A}^T\mathbf{P} + \mathbf{P}\mathbf{A}\right)\mathbf{x} + 2\mathbf{x}^T\mathbf{P}\mathbf{b}w_x(L,t)\right). \tag{4.134}$$

The Meyer-Kalman-Yakubovich Lemma [118] ensures that for $G(s)$ satisfying Eq. (4.122) we have for all $\mathbf{L} = \mathbf{L}^T > 0$, there exist $\mathbf{P} = \mathbf{P}^T > 0$, $\varepsilon > 0$, and $\mathbf{q}$ such that

$$\mathbf{A}^T\mathbf{P} + \mathbf{P}\mathbf{A} = -\mathbf{q}\mathbf{q}^T - \varepsilon\mathbf{L}, \tag{4.135}$$

$$\mathbf{P}\mathbf{b} = \frac{1}{2}\mathbf{c} + \mathbf{q}\sqrt{d-\gamma}. \tag{4.136}$$

Equation (4.136) requires $d > \gamma > 0$, implying that $G(s)$ must have relative degree zero. Substitution of Eqs. (4.132), (4.135), and (4.136) into Eq. (4.134) yields

$$\dot{V}_z = -P\left\{w_t(L,t)w_x(L,t) + \varepsilon\mathbf{x}^T\mathbf{L}\mathbf{x} + \gamma w_x^2(L,t) + \left[\mathbf{x}^T\mathbf{q} - w_x(L,t)\sqrt{d-\gamma}\right]^2\right\}. \tag{4.137}$$

Lemma 32 *The Lyapunov functional* $V = E + 2C + V_z$ *with*

$$\beta \leq \frac{1}{4L}\min\{\rho, P\} = \beta_1 \tag{4.138}$$

is lower and upper bounded as follows:

$$\underline{\kappa}\ ||\mathbf{z}||^2 \leq V \leq \overline{\kappa}\ ||\mathbf{z}||^2, \tag{4.139}$$

where $\underline{\kappa}$, $\overline{\kappa} > 0$, *the state vector*

$$\mathbf{z} = \left[w_t(x,t),\ w_x(x,t),\ \mathbf{x}(t)^T\right]^T \qquad x \in [0, L], \tag{4.140}$$

and the norm

$$||\mathbf{z}||^2 = \int_0^L \left(w_t^2 + w_x^2\right) dx + ||\mathbf{x}||_e^2, \tag{4.141}$$

where $||\cdot||_e$ *is the standard Euclidean norm.*

Proof: Using Ineq. (3.10), C is upper bounded as follows

$$\begin{aligned} C &= \beta \int_0^L x w_t w_x dx \leq \beta \int_0^L L\left(w_t^2 + w_x^2\right) dx \\ &\leq \frac{\beta L}{\min\{\rho, P\}} \int_0^L \left(\rho w_t^2 + P w_x^2\right) dx = \frac{2\beta L}{\min\{\rho, P\}} E. \end{aligned} \tag{4.142}$$

Inequality (4.142) implies that

$$-\frac{2\beta L}{\min\{\rho, P\}}\ E \leq C \leq \frac{2\beta L}{\min\{\rho, P\}}\ E. \tag{4.143}$$

Addition of E and V_z to twice Ineq. (4.143) yields

$$E - \frac{4\beta L}{\min\{\rho, P\}}\ E + V_z \leq V \leq E + \frac{4\beta L}{\min\{\rho, P\}}\ E + V_z. \tag{4.144}$$

Finally,

$$\underline{\kappa} = \min\left\{\frac{1}{2}\left(\min\{\rho, P\} - 4\beta L\right), P\lambda_{\min}(\mathbf{P})\right\} \tag{4.145}$$

and

$$\overline{\kappa} = \max\left\{\frac{1}{2}\left(\max\{\rho, P\} - \frac{4\beta L}{\min\{\rho, P\}}\right), P\lambda_{\max}(\mathbf{P})\right\}. \tag{4.146}$$

Note that $\underline{\kappa} > 0$ for β satisfying Inequality (4.138).

□

Lemma 33 *For*

$$\beta \le \min\left\{ \frac{\varepsilon P\lambda_{\min}(\mathbf{L})}{L\left(||\mathbf{c}||_e^2 + 2||\mathbf{c}||_e\ |d|\right)},\ \frac{P\gamma}{L\left(\frac{P}{\rho} + d^2 + 2||\mathbf{c}||_e\ |d|\right)} \right\} = \beta_2 \tag{4.147}$$

there exists $\kappa > 0$ *such that*

$$\dot{V} \le -\kappa\ ||\mathbf{z}||^2. \tag{4.148}$$

Proof: The time derivative of the Lyapunov functional,

$$\begin{aligned}\dot{V} &= \beta L w_t^2(L,t) - P\left(\gamma - \frac{\beta L}{\rho}\right) w_x^2(L,t) - \beta \int_0^L \left(w_t^2 + \frac{P}{\rho} w_x^2\right) dx - \varepsilon P \mathbf{x}^T \mathbf{L}\mathbf{x} \\ &\quad - P\left[\mathbf{x}^T\mathbf{q} - w_x(L,t)\sqrt{d-\gamma}\right]^2 \end{aligned}\tag{4.149}$$

can be upper bounded by:

$$\dot{V} \le \beta L w_t^2(L,t) - P\left(\gamma - \frac{\beta L}{\rho}\right) w_x^2(L,t) - \beta \int_0^L \left(w_t^2 + \frac{P}{\rho} w_x^2\right) dx - \varepsilon P \mathbf{x}^T \mathbf{L}\mathbf{x}. \tag{4.150}$$

Furthermore,

$$\dot{V} \le \beta L w_t^2(L) - P\left(\gamma - \frac{\beta L}{\rho}\right) w_x^2(L) - \beta \int_0^L \left(w_t^2 + \frac{P}{\rho} w_x^2\right) dx - \varepsilon P \lambda_{\min}(L)||\mathbf{x}||_e^2 \tag{4.151}$$

because $\mathbf{L} = \mathbf{L}^T > 0$. Using the output Eq. (4.132), the first term in Eq. (4.151) can be rewritten and upper bounded as follows:

$$\begin{aligned} w_t^2(L,t) &= \left[\mathbf{c}^T\mathbf{x} + d w_x(L)\right]^2 \\ &\le ||\mathbf{c}||_e^2||\mathbf{x}||_e^2 + d^2 w_x^2(L,t) + 2\ ||\mathbf{c}||_e\ ||\mathbf{x}||_e\ |d|\ |w_x(L,t)|. \end{aligned}\tag{4.152}$$

Inequality (3.10) implies that:

$$w_t^2(L) \le ||\mathbf{c}||_e^2||\mathbf{x}||_e^2 + d^2 w_x^2(L,t) + 2||\mathbf{c}||_e\ |d|\ \left[||\mathbf{x}||_e^2 + w_x^2(L,t)\right]. \tag{4.153}$$

Substitution of Ineq. (4.153) into Ineq. (4.151) yields:

$$\begin{aligned} \dot{V} &\le -\beta \int_0^L \left(w_t^2 + \frac{P}{\rho} w_x^2\right) dx \\ &\quad - \left[\varepsilon P \lambda_{\min}(L) - \beta L\left(||\mathbf{c}||_e^2 + 2||\mathbf{c}||_e\ |d|\right)\right] ||\mathbf{x}||^2 \\ &\quad - \left[P\gamma - \beta L\left(\frac{P}{\rho} + d^2 + 2||\mathbf{c}||_e\ |d|\right)\right] w_x^2(L) \qquad (4.154) \\ &\le -\beta \int_0^L \left(w_t^2 + \frac{P}{\rho} w_x^2\right) dx \\ &\quad - \left[\varepsilon P \lambda_{\min}(L) - \beta L\left(||\mathbf{c}||_e^2 + 2||\mathbf{c}||_e\ |d|\right)\right] ||\mathbf{x}||^2. \qquad (4.155) \end{aligned}$$

Choosing β as in Ineq. (4.147) ensures that the two bracketed terms in Ineq. (4.154) are positive. Finally, from Ineq. (4.155) and the definition of the norm (4.141),

$$\kappa = \min\left\{\beta,\ \beta\frac{P}{\rho},\ \varepsilon P\lambda_{\min}(L) - \beta L\left(||\mathbf{c}||_e^2 + 2||\mathbf{c}||_e\ |d|\right)\right\} > 0, \tag{4.156}$$

and Eq. (4.148) holds.
□

Lemma 34 *For*

$$\beta \ \leq\ \min\left\{\beta_1,\ \beta_2\right\}, \tag{4.157}$$

there exists $K > 0$ *such that:*

$$V(t)\ \leq\ V(0)\ e^{-Kt}. \tag{4.158}$$

Proof: Lemma 32 and Lemma 33 produce

$$\dot{V} \leq\ -\frac{\kappa}{\overline{\kappa}}V. \tag{4.159}$$

Definition of $K = \kappa/\overline{\kappa}$ and integration of Eq. (4.159) yield Eq. (4.158).
□

Using Ineq. (3.5) and Lemma 3,

$$\frac{P}{2L}w^2(x,t)\ \leq\ \frac{P}{2}\int_0^L w_x^2(t)dx\ \leq\ V(t)\ \leq\ V(0)\ \mathrm{e}^{-Kt}, \tag{4.160}$$

providing the strong (pointwise) exponential stability result:

$$w(x,t)\ \leq\ \frac{\sqrt{2V(0)}}{c}\ \mathrm{e}^{-\frac{K}{2}t} \qquad \forall\ x \in [0,L]\,.$$

□

Control Design Equation (4.120) defines a general class of boundary controllers. Example $H_1(s)$ and $H_2(s)$ can have proportional, integral, and derivative terms as follows:

$$H_1(s) \ =\ c_P + c_I\frac{1}{s} + c_D s, \tag{4.161}$$

$$H_2(s) \ =\ k_D + k_P\frac{1}{s}, \tag{4.162}$$

where $c_P,\ c_I,\ c_D,\ k_P,\ k_D$ are positive control gains. Substitution of Eqs. (4.161) and (4.162) into Eq. (4.121) produces:

$$G(s) = \frac{c_D s^2 + (c_P + P)\, s + c_I}{m s^2 + k_D s + k_P}. \tag{4.163}$$

The transfer function (4.163) has relative degree zero if $c_D \neq 0$. The corresponding boundary control law is:

$$f(t) = -\left(c_P w_x(L) + c_I \int_0^L w_x(L,\tau) d\tau + c_D w_{xt}(L) + k_P w(L) + k_D w_t(L) \right). \tag{4.164}$$

The controller gains are chosen such that the transfer function $G(s)$ satisfies Eq. (4.122).

Experimental Results

Test Stand Description Using the experimental test stand shown in Figure 1, we implement the dynamic compensator in Eq. (4.164). In this experiment, a braided polyester rope with $L = 4.038$ m, $\rho = 0.092$ kg/m, $P = 3.1$ N, and $m = 1.5$ kg is used. A Hohner 1000-count rotary encoder is added to measure the string departure angle $w_x(L,t)$. Filtered backwards differencing and trapezoidal integration provide gantry and departure angular velocities and the departure angle integral, respectively.

Controller Design Pole placement of $G(s)$ at $-1.67 \pm 3.25j$ yields $k_P = 20$, and $k_D = 5$, ensuring $G(s)$ is strictly stable. The remaining three control gains ($c_P = 200$, $c_I = 5$, and $c_D = 10$) are chosen to ensure that the $G(j\omega)$ phase magnitude ($|\angle G(j\omega)|$) is less than 90° for all ω. Consequently, $G(s)$ satisfies Eq. (4.122).

Experimental Results At $t = 3$ sec. a 20 Ns impulse disturbance is applied to the string mid-span using a Piezotronics PCB 086 C03 impulse hammer. Figure 5 shows the open and closed loop system response. Frames (a) - (c) show the gantry position, the vibrating departure angle, and the motor voltage, respectively. The dotted lines show the open loop response decaying slowly under natural damping. The solid lines show the performance of the boundary controller (4.164) using the calculated gains. The closed loop response decays in approximately 2 seconds, more than six times faster than the open loop response. In addition, the maximum angular response is reduced by more than 50%.

In the experiment, the end point displacement and departure angle decay rapidly and in an apparently exponential manner, verifying the theorem. High frequency activity in the departure angle and control voltage signals indicates damping of high frequency modes. Thus the experimental results provide support for the theoretically proven exponential stability of the distributed model.

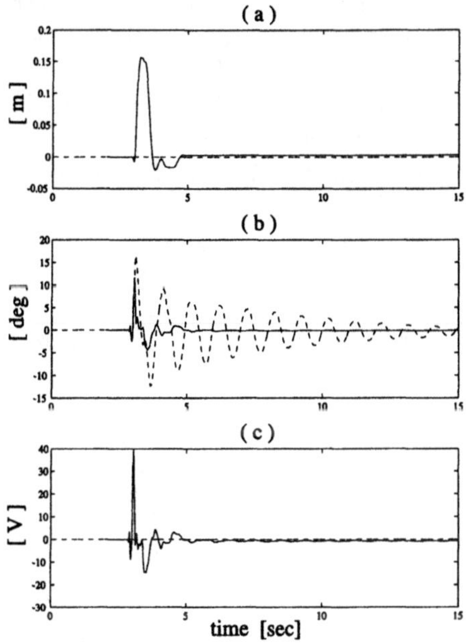

Figure 5 Dynamic boundary controller experimental impulse response: – – – Open Loop; —— Closed Loop (a) End–Point Displacement, $w(L,t)$; (b) End–Point Departure Angle, $w_x(L,t)$; (c) Control voltage

Accelerometer Feedback The theory developed in the previous section can be applied to accelerometer feedback. The control law (4.164) includes filters $H_1(s)$ and $H_2(s)$ on the velocity and angle signals, respectively. It is often more cost effective to use an accelerometer to measure boundary velocity. Typically, the accelerometer signal is filtered by the bandpass filter

$$G_{bp}(s) = \frac{\tau^2 s}{(\tau s + 1)^2}, \tag{4.165}$$

where τ is the filter time constant. Figures 6(a) and (b) show the frequency response of G_{bp}. At low frequency G_{bp} acts as a high pass filter to eliminate the drift characteristic of piezoelectric accelerometers. At the cutoff frequency, $\omega = 1/\tau$, the magnitude curve turns and converges to the integrator $(1/\omega)$ line. The phase initially leads by 90° and eventually falls to the integrator delay of -90°.

Thus the transfer function from $V(s)$ to $F(s)$ is the product of the bandpass filter $G_{bp}(s)$ times s, or

$$H_2(s) = \frac{\tau^2 s^2}{(\tau s + 1)^2}, \tag{4.166}$$

assuming the accelerometer exactly measures $w_{tt}(L,t)$. If we choose

$$H_1(s) = ms - P, \tag{4.167}$$

then

$$G(s) = \frac{m(\tau s + 1)^2}{m\tau^2 s^2 + \tau(2m + \tau)s + m}. \tag{4.168}$$

Figures 6(c) and (d) show the frequency response magnitude notching around the cutoff frequency. The phase remains within ±90°, satisfying the condition (4.122).

4.3 Free Boundary Problems

4.3.1 Gantry Crane

Now we return to the gantry crane example and use LaSalle's invariance principle to show asymptotic stability under the proposed control law. First, we establish precompactness of solution trajectories.

Lemma 35 *For the operator $\mathcal{A}$ given in Eq.(3.39), $0 \in \mathcal{R}(\mathcal{A})$ and $(\lambda\mathcal{I} - \mathcal{A})^{-1}$ is compact for some $\lambda > 0$.*

Proof: It is obvious that $0 \in \mathcal{R}(\mathcal{A})$. For the compactness of $(\lambda\mathcal{I} - \mathcal{A})^{-1}$, we first show that $\mathcal{A}^{-1}$ exists and maps any bounded set of $\mathcal{H}$ into a bounded set of

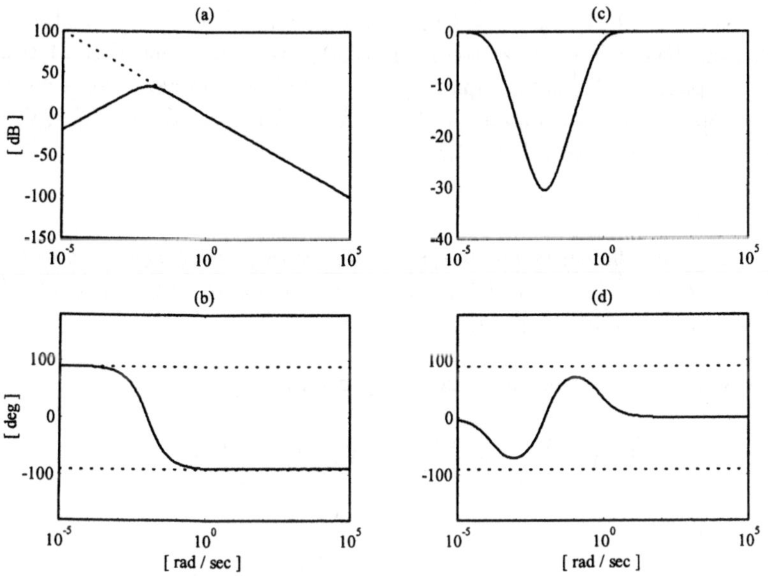

Figure 6 Bode plots of the accelerometer filter $H_2(s)$ magnitude (a) and phase (b) and overall transfer function $G(s)$ magnitude (c) and phase (d).

$\mathcal{D}(\mathcal{A})$. Let $\widetilde{\mathbf{z}} = [\widetilde{u}, \widetilde{\nu}, \widetilde{\varphi}, \widetilde{\psi}, \widetilde{\vartheta}, \widetilde{\varrho}]^T \in \mathcal{H}$, by solving $\mathcal{A}\mathbf{z} = \widetilde{\mathbf{z}}$ for $\mathbf{z} = [u, \nu, \varphi, \psi, \vartheta, \varrho]^T$, we have

$$\begin{array}{ll} u = \int_0^x \frac{1}{p_o - \sigma} \left(\int_0^\sigma \widetilde{\nu} d\xi + c_1 \right) d\sigma + c_2, & \text{(a)} \\ v = \widetilde{u}, & \text{(b)} \\ \psi = \widetilde{\varphi}, & \text{(c)} \\ -k_p \varphi - k_d \psi + k_a u_x(0) = \widetilde{\psi}, & \text{(d)} \\ \varrho = \widetilde{\vartheta}, & \text{(e)} \\ -u_x(1) = \widetilde{\varrho}, & \text{(f)} \end{array} \tag{4.169}$$

where c_1 and c_2 can be uniquely determined from (4.169f) and (4.169d), respectively. Therefore, $\mathcal{A}^{-1}$ exists and $\mathbf{z} = [u, \nu, \varphi, \psi, \vartheta, \varrho]^T \in \mathcal{D}(\mathcal{A})$. Moreover, if $\widetilde{\mathbf{z}}$ is bounded in $\mathcal{H}$, $\mathbf{z}$ is bounded in $\mathcal{D}(\mathcal{A})$. Since $\mathcal{A}^{-1}$ maps any bounded set of $\mathcal{H}$ into a bounded set of $\mathcal{D}(\mathcal{A})$, from Rellich's Lemma (see p.14 [123]), the embedding of the latter space into $\mathcal{H}$ is compact, so $\mathcal{A}^{-1}$ is a compact operator. This also proves that $\mathcal{A}$ has isolated eigenvalues and that if we choose λ different from the eigenvalues of $\mathcal{A}$, $(\lambda\mathcal{I} - \mathcal{A})^{-1}$ is a compact operator (see p.187 [61]).

□

Theorem 36 *The displacement $w(x,t)$ of the gantry crane system given by Eqs. (2.21) - (2.23) under the control law Eq. (2.99) decays to zero asymptotically with time for any $\mathbf{z}_0 \in \mathcal{H}$ and $x \in [0,1]$.*

Proof: First, we calculate the energy from Eq. (2.105)

$$E(t) = \frac{1}{2} \int_0^1 w_t^2 + (p_o - x)^2 w_x^2 dx + \frac{p_0^2 k_p}{2k_a} w^2(0)$$

with time derivative

$$\dot{E} = -k_d w_t^2(0) \leq 0.$$

To prove the theorem, we use show that, for an initial data $\mathbf{z}_0 = [u_0, \nu_0, \varphi_0, \psi_0, \vartheta_0, \varrho_0]^T \in \mathcal{D}(\mathcal{A})$, the condition $\dot{E}(t) \equiv 0$ implies $\mathbf{z}_0 = [0,0,0,0,0,0]^T$. Lemma 19 and Theorem 20 extend the result to any $\mathbf{z}_0 \in \mathcal{H}$. From $\dot{E}(t) \equiv 0$, we know that $w_t(0,t) \equiv 0$, so $\mathbf{z}(t) = [u, \nu, \varphi, \psi, \vartheta, \varrho]^T$ satisfies

$$\begin{array}{ll} w_{tt} - [(p_o - x) w_x]_x = 0, & \text{(a)} \\ w_t(0) = 0, & \text{(b)} \\ k_p w(0) - k_a w_x(0) = 0, & \text{(c)} \\ w_{tt}(1) + w_x(1) - 0. & \text{(d)} \end{array} \tag{4.170}$$

Since all the boundary conditions are separable, the solution of (4.170) can be found by using separation of variables [81] as follows:

$$w(x,t) = W(x)\eta(t). \tag{4.171}$$

We distinguish two cases that produce $w_t(0,t) \equiv 0$:

i) $\dot{\eta}(t) \equiv 0$. From Eq. (4.170a), it follows that $(p_o - x)W_x(x) = c_1$, where c_1 is an arbitrary constant. From Eq. (4.170d), however, we obtain $c_1 = 0$, so $W(x) = c_2$. Eq. (4.170c) requires $c_2 = 0$, yielding $w(x,t) = 0$.

ii) $W(0) = 0$. The eigenvalue problem generated by substitution of Eq. (4.171) into (4.170) with $\eta(t) = e^{i\omega t}$ is

$$\omega^2 W + [(p_o - x)W_x]_x = 0. \tag{4.172}$$

As in Chapter 2, the variable definition

$$s = 2\omega\sqrt{p_o - x} \tag{4.173}$$

transforms Eq. (4.172) to

$$s\psi + \psi_s + s\psi_{ss} = 0 \qquad \text{for } s \in (s_0, s_1), \tag{4.174}$$

where

$$s_0 = 2\omega\sqrt{p_o}, \quad s_1 = 2\omega\sqrt{p_o - 1}, \tag{4.175}$$

and $\psi(s) = W(x)$. The solutions of Eq. (4.174) are

$$\psi(s) = C_1 J_0(s) + C_2 Y_0(s), \tag{4.176}$$

where J_0 and Y_0 are Bessel functions of the first and second kind, respectively [1] and C_1, C_2 and ω are to be determined. From (4.170b) and (4.170c), we have $W(0) = 0$ and $W_x(0) = 0$, yielding the two equations

$$\begin{bmatrix} J_0(s_0) & Y_0(s_0) \\ J_1(s_0) & Y_1(s_0) \end{bmatrix} \begin{bmatrix} C_1 \\ C_2 \end{bmatrix} = 0.$$

Now we show that the determinant of this matrix is non zero for all $s_0 > 0$. Let

$$\Phi(s_0) = \det \begin{bmatrix} J_0(s_0) & Y_0(s_0) \\ J_1(s_0) & Y_1(s_0) \end{bmatrix} = J_0(s_0)Y_1(s_0) - J_1(s_0)Y_0(s_0).$$

From the properties of Bessel functions [1], we know that

$$\Phi(1) \neq 0, \tag{4.177}$$

and

$$\frac{d}{ds_0}\Phi(s_0) = -\frac{\Phi(s_0)}{s_0} \qquad \forall s_0 > 0. \tag{4.178}$$

Now suppose there exists some $s_{01} > 0$ such that $\Phi(s_{01}) = 0$. From (4.178), however, this implies $\Phi(s_0) = 0$ for all $s_0 > 0$, in contradiction with Eq. (4.177). Therefore, $\Phi(s_0) \neq 0$ for all $s_0 > 0$, so $C_1 = C_2 = 0$, yielding $w(x,t) = 0$. □

4.3.2 Flexible Link Robot

We assume that for the flexible link robot governed by the matrix operator Eq. (2.110) with inner product Eq. (2.111) there exists a unique solution and that solution trajectories are precompact. We now show that $w(x,t) = 0$ is asymptotically stable with PD feedback control

$$\tau_h(t) = -k_p w_x(0) - k_d w_{xt}(0), \tag{4.179}$$

where k_p and k_d are positive control gains. The system energy

$$\begin{aligned} E &= \frac{1}{2}\int_0^L \rho w_t^2 + EI w_{xx}^2 dx + \frac{1}{2} J_h w_{xt}^2(0) \\ &\quad + \frac{1}{2} k_p w_x^2(0) + \frac{1}{2} m w_t^2(L) + \frac{1}{2} J_m w_{xt}^2(L) \end{aligned} \tag{4.180}$$

has a time derivative

$$\dot{E} = -k_d w_{xt}^2(0).$$

According to LaSalle's Invariance Principle, the origin is asymptotically stable if $\dot{E} \equiv 0$ implies $w(x,t) \equiv 0$. With $\dot{E} \equiv 0$ we have $w_{xt}(0) \equiv 0$ and $w_{xtt}(0) \equiv 0$, so the boundary conditions (2.51) - (2.54) simplify to

$$w(0) = 0 \tag{4.181}$$

$$k_p w_x(0) - EI w_{xx}(0) = 0 \tag{4.182}$$

$$w_{tt}(L) - \frac{EI}{m} w_{xxx}(L) = 0 \tag{4.183}$$

$$w_{xtt}(L) + \frac{EI}{J_m} w_{xx}(L) = 0 \tag{4.184}$$

The field equations and boundary conditions are separable so we substitute Eq. (4.171) into Eqs. (4.181) - (4.184) and the field equation (2.50) with $\eta(t) = e^{i\omega t}$ to obtain an eigenvalue problem with solutions of the form

$$W(x) = C_1 \sin(\beta x) + C_2 \cos(\beta x) + C_3 \sinh(\beta x) + C_4 \cosh(\beta x). \tag{4.185}$$

For $\dot{\eta} \neq 0$, the condition $w_{xt}(0) \equiv 0$ implies

$$W_x(0) = C_1 + C_3 = 0$$

Substitution of Eq. (4.185) into Eqs. (4.181) and (4.182) with $W_x(0) = 0$ yields

$$\begin{aligned} W(0) &= C_2 + C_4 = 0, \\ W_{xx}(0) &= -C_2 + C_4 = 0, \end{aligned}$$

or $C_2 = C_4 = 0$. Thus, Eq. (4.185) simplifies to

$$W(x) = C_1 (\sin(\beta x) - \sinh(\beta x)).$$

and boundary condition (4.183)

$$\omega^2 W(L) + \frac{EI}{m} W_{xxx}(L) = 0,$$

becomes

$$\left(\omega^2 (\sin(\beta L) - \sinh(\beta L)) - \frac{EI}{m}\beta^3 (\cos(\beta L) + \cosh(\beta L))\right) C_1 = 0. \tag{4.186}$$

Substituting Eq. (2.151), we obtain

$$\frac{\beta m}{\rho} [\sin(\beta L) - \sinh(\beta L)] + [\cos(\beta L) - \cosh(\beta L)] = 0 \tag{4.187}$$

for a nonzero ($C_1 \neq 0$) $w(x, t)$. Each of the two bracketed terms in Eq. (4.187), however, are negative for all βL. Thus Eq. (4.187) has no solution and $C_1 = 0$ in Eq. (4.186). This means that the only possible solution with $\dot{E} \equiv 0$ is the origin ($C_1 = C_2 = C_3 = C_4 = 0$) and the closed loop system is asymptotically stable.

Chapter 5
EXACT MODEL KNOWLEDGE CONTROL

In high performance applications, passive control may not provide sufficient vibration or noise control. This chapter introduces active, model-based controllers that damp vibration and noise. Exact knowledge of the model structure and parameters provides controllers with fast transient decay and excellent disturbance rejection. Exact model knowledge (EMK) controllers are introduced that compensate for the effects of actuator dynamics, geometric nonlinearities, and spatial parameter nonuniformity.

For EMK boundary control, the boundaries must either be constrained or controlled. In this chapter, we apply the EMK approach to a pinned-controlled string and cable, an acoustic duct with a boundary control speaker, and a boundary controlled cantilevered beam. With one control input on each boundary, we can translate the material and damp vibration. We demonstrate speed, tension, and vibration control for axially moving materials using this approach.

Two control inputs can be colocated at the coupling point of two subdomains to exponentially stabilize each subdomain. If two actuators are not feasible, domain control using a single actuator can isolate a "quiet" subdomain from bounded disturbances in the other "noisy" subdomain. This technique is demonstrated for acoustic noise isolation and axially moving beam transverse vibration isolation.

High speed video and shaped piezoelectric laminates can provide distributed measurements. In the last section of this chapter we show that an EMK modal controller can be developed for distributed systems based on distributed measurements. The technique is experimentally demonstrated on a single flexible link robot arm.

5.1 Boundary Control

With colocated boundary measurements of displacement, velocity and slope for second order systems, boundary controllers use force and torque actuators to provide strong exponential stability. We can obtain similar results for fourth order system with additional curvature and shear measurements.

5.1.1 Second Order Systems

In this section, we develop an EMK boundary controller for the second order acoustic duct model. The governing equations are similar to the boundary controlled string Eqs. (2.14) - (2.16), where, unlike in Chapter 4, the boundary condition in Eq. (2.16)

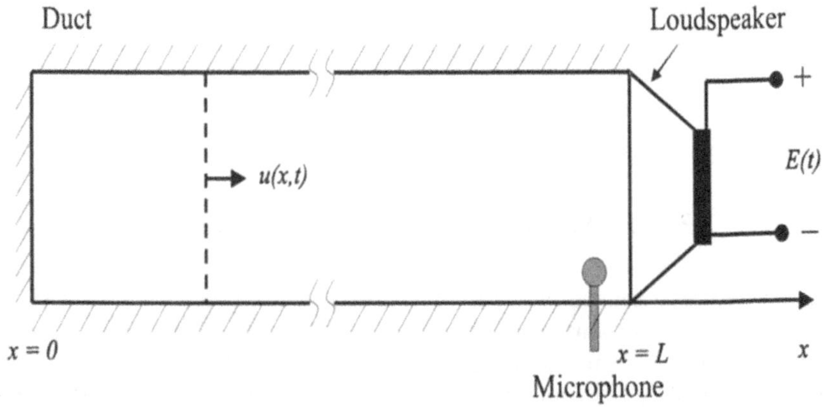

Figure 1 Schematic diagram of the boundary controlled duct.

includes the actuator inertial dynamics ($m \neq 0$). The control objective is to exponentially stabilize the acoustic displacement using the boundary force control input.

Boundary Control of an Acoustic Duct

Figure 1 shows a schematic diagram of the finite duct, sealed at one end and actuated at the other by a loudspeaker. The field equation and boundary conditions are

$$\begin{aligned} \rho A u_{tt} - \mathcal{B} A u_{xx} &= 0 \text{ for } x \in (0, L), && (5.1)\\ u(0,t) &= 0, && (5.2)\\ m u_{tt}(L,t) + \mathcal{B} A u_x(L,t) + B_d u_t(L,t) + K_s u(L,t) &= f(t) && (5.3) \end{aligned}$$

where B_d and K_s are the damping and stiffness of loudspeaker actuator, respectively. We neglect the speaker electrical dynamics by assuming that the control force is proportional to the applied voltage $E(t)$.

In Chapter 4, we show that a boundary damper exponentially stabilizes the system governed by Eqs. (5.1) - (5.3) with $m = B_d = K_s = 0$. The EMK control law introduced in this section is based on the concept of backstepping. Specifically, we design the control law to drive an auxiliary signal

$$\eta(t) = u_t(L,t) + \alpha u_x(L,t) \qquad (5.4)$$

to zero, where α is a positive scalar gain. If the controller successfully drives $\eta \to 0$, then the damper boundary condition of Eq. (2.15) is enforced and an exponential decay of the string displacement can be achieved.

After taking the time derivative of Eq. (5.4), multiplying throughout by m, and substituting for $m u_{tt}(L,t)$ from Eq. (5.3), we obtain the open-loop dynamics for

$\eta(t)$

$$m\dot{\eta}(t) = m\alpha u_{xt}(L,t) - \mathcal{B}Au_x(L,t) - B_d u_t(L,t) - K_s u(L,t) + f(t). \tag{5.5}$$

The EMK boundary control force is defined as

$$f(t) = -m\alpha u_{xt}(L,t) + \mathcal{B}Au_x(L,t) + B_d u_t(L,t) + K_s u(L,t) - \kappa\eta(t), \tag{5.6}$$

where κ is a positive, scalar, constant feedback gain. Substitution of Eq. (5.6) into Eq. (5.5), we obtain the closed-loop boundary dynamics

$$m\dot{\eta}(t) = -\kappa\eta(t). \tag{5.7}$$

Theorem 37 *The EMK boundary control law given by Eq. (5.6) strongly exponentially stabilizes the acoustic displacement.*

Proof: We first define the Lyapunov functional

$$V(t) = E_s(t) + E_c(t) + \frac{1}{2}m\eta^2(t), \tag{5.8}$$

where the acoustic energy is

$$E_s = \frac{1}{2}\beta_s\rho A\int_0^L u_t^2 dx + \frac{1}{2}\beta_s\mathcal{B}A\int_0^L u_x^2 dx, \tag{5.9}$$

the crossing term is

$$E_c = 2\beta_c\rho A\int_0^L xu_t u_x dx, \tag{5.10}$$

where β_s and β_c are positive, scalar constants. Note that the design constants β_c and β_s do not appear in the control law of Eq. (5.6).

Lemma 38 *The function given by Eq. (5.8) is bounded as follows*

$$\lambda_1\left[E_s(t) + \eta^2(t)\right] \le V(t) \le \lambda_2\left[E_s(t) + \eta^2(t)\right], \tag{5.11}$$

where the positive scalar constants

$$\begin{aligned}\lambda_1 &= \min\left\{1 - \frac{4\beta_c\rho L}{\beta_s\min\{\rho,\mathcal{B}\}}, \frac{m}{2}\right\}, \\ \lambda_2 &= \max\left\{1 + \frac{4\beta_c\rho L}{\beta_s\min\{\rho,\mathcal{B}\}}, \frac{m}{2}\right\}.\end{aligned} \tag{5.12}$$

Proof: The acoustic energy is lower bounded by

$$E_s \geq \beta_s A \min\{\rho, \mathcal{B}\}\left[\frac{1}{2}\int_0^L \left(u_t^2 + u_x^2\right) dx\right] \tag{5.13}$$

and the crossing term is upper bounded as

$$E_c \leq 4\beta_c \rho AL\left[\frac{1}{2}\int_0^L \left(u_t^2 + u_x^2\right) dx\right], \tag{5.14}$$

where Ineq. (3.10) has been used. We now use Ineq. (5.13) to obtain the following bounds for $E_c(t)$

$$-\left(\frac{4\beta_c \rho L}{\beta_s \min\{\rho, \mathcal{B}\}}\right) E_s \leq E_c \leq \left(\frac{4\beta_c \rho L}{\beta_s \min\{\rho, \mathcal{B}\}}\right) E_s. \tag{5.15}$$

Hence,

$$0 \leq \xi_1 E_s \leq E_s + E_c \leq \xi_2 E_s, \tag{5.16}$$

where the scalar constants

$$\xi_1 = 1 - \frac{4\beta_c \rho L}{\beta_s \min\{\rho, \mathcal{B}\}} \qquad \xi_2 = 1 + \frac{4\beta_c \rho L}{\beta_s \min\{\rho, \mathcal{B}\}} \tag{5.17}$$

are positive if

$$\beta_c < \frac{\beta_s \min\{\rho, \mathcal{B}\}}{4\rho L}. \tag{5.18}$$

The bound given in Eq. (5.11) results from the structure of Eqs. (5.8) and (5.16). □

Lemma 39 *The time derivative of the Lyapunov functional $V(t)$ defined in Eq. (5.8) is upper bounded as follows*

$$\dot{V}(t) \leq -\lambda_3\left(\eta^2(t) + E_s(t)\right), \tag{5.19}$$

where

$$\lambda_3 = \min\left\{\left(\kappa - \frac{\beta_s \mathcal{B} A}{2\alpha}\right), 2\frac{\beta_c}{\beta_s}\right\} > 0. \tag{5.20}$$

Proof: After taking the time derivative of Eq. (5.8), and substituting for $m\dot{\eta}(t)$ from Eq. (5.7), we obtain

$$\dot{V} = -\kappa\eta^2 + \dot{E}_s + \dot{E}_c. \tag{5.21}$$

To determine $\dot{E}_s(t)$ in Eq. (5.21), we take the time derivative of Eq. (5.9), substitute the field equation (5.1), and cancel common terms to obtain

$$\dot{E}_s = \left[\beta_s \mathcal{B} A \int_0^L u_{xx} u_t dx\right] + \beta_s \mathcal{B} A \int_0^L u_x u_{xt} dx. \tag{5.22}$$

After integrating the bracketed term in Eq. (5.22) by parts, substituting the boundary condition of Eq. (5.2), and then cancelling common terms, we obtain

$$\dot{E}_s = \beta_s \mathcal{B} A u_x(L) u_t(L), \tag{5.23}$$

which is rewritten as

$$\dot{E}_s = -\frac{\beta_s \mathcal{B} A}{2\alpha} \left[\alpha^2 u_x^2(L) + u_t^2(L)\right] + \frac{\beta_s \mathcal{B} A}{2\alpha} \eta^2. \tag{5.24}$$

To determine $\dot{E}_c(t)$ in Eq. (5.21), we take the time derivative of Eq. (5.10) to obtain

$$\dot{E}_c = 2\beta_c \rho A \int_0^L x u_t u_{xt} dx + 2\beta_c \rho A \int_0^L x u_{tt} u_x dx. \tag{5.25}$$

After substituting for $\rho u_{tt}(x,t)$ from Eq. (5.1), we obtain

$$\dot{E}_c = A_1 + A_2, \tag{5.26}$$

where

$$A_1 = 2\beta_c \rho A \int_0^L x u_t u_{xt} dx \qquad A_2 = 2\beta_c \mathcal{B} A \int_0^L x u_{xx} u_x dx. \tag{5.27}$$

After integrating A_1 in Eq. (5.27) by parts, we obtain

$$A_1 = 2\beta_c \rho A L u_t^2(L) - 2\beta_c \rho A \int_0^L u_t^2 dx - \left[2\beta_c \rho A \int_0^L x u_t u_{xt} dx\right], \tag{5.28}$$

where Eq. (5.2) has been used. After observing that the bracketed term in Eq. (5.28) is equal to A_1, defined in Eq. (5.27), we rewrite Eq. (5.28) as follows

$$A_1 = \beta_c \rho A L u_t^2(L) - \beta_c \rho A \int_0^L u_t^2 dx. \tag{5.29}$$

After integrating A_2 in Eq. 5.27) by parts, we obtain

$$A_2 = 2\beta_c \mathcal{B} A L u_x^2(L) - 2\beta_c \mathcal{B} A \int_0^L u_x^2 dx - \left[2\beta_c \mathcal{B} A \int_0^L x u_{xx} u_x dx\right], \tag{5.30}$$

where Eq. (5.2) is used. After observing that the bracketed term in Eq. (5.30) is equal to A_2, defined in Eq. (5.27), we rewrite Eq. (5.30) as follows

$$A_2 = \beta_c \mathcal{B} A L u_x^2(L) - \beta_c \mathcal{B} A \int_0^L u_x^2 dx. \tag{5.31}$$

After substituting Eqs. (5.29) and (5.31) into Eq. (5.26), we obtain

$$\dot{E}_c = \beta_c \rho A L u_t^2(L) + \beta_c \mathcal{B} A L u_x^2(L) - \beta_c \rho A \int_0^L u_t^2 dx - \beta_c \mathcal{B} A \int_0^L u_x^2 dx. \tag{5.32}$$

After substituting Eqs. (5.32) and (5.24) into (5.21), and appropriately rearranging terms, we obtain

$$\begin{aligned}\dot{V} &= -\left(\kappa-\frac{\beta_s\mathcal{B}A}{2\alpha}\right)\eta^2-\left(\frac{\beta_s\mathcal{B}A}{2\alpha}-\beta_c\rho AL\right)u_t^2(L)-\left(\frac{\beta_s\mathcal{B}A\alpha}{2}-\beta_c\mathcal{B}AL\right)u_x^2(L)\\ &\quad -\beta_c\rho A\int_0^L u_t^2dx-\beta_c\mathcal{B}A\int_0^L u_x^2dx. \end{aligned} \tag{5.33}$$

If β_c and β_s are selected to satisfy the following inequalities

$$\beta_c<\min\left\{\frac{\beta_s\mathcal{B}}{2\rho L\alpha},\frac{\beta_s\alpha}{2L}\right\},\quad \beta_s<\frac{2\kappa\alpha}{\mathcal{B}A}, \tag{5.34}$$

we can upper bound Eq. (5.33) as follows

$$\dot{V}\leq-\left(\kappa-\frac{\beta_s\mathcal{B}A}{2\alpha}\right)\eta^2-2\frac{\beta_c}{\beta_s}\left(\frac{1}{2}\beta_s\rho A\int_0^L u_t^2dx\right)-2\frac{\beta_c}{\beta_s}\left(\frac{1}{2}\beta_s\mathcal{B}A\int_0^L u_x^2dx\right) \tag{5.35}$$

After substituting Eq. (5.9) into Eq. (5.35), we obtain

$$\dot{V}\leq-\min\left\{\left(\kappa-\frac{\beta_s\mathcal{B}A}{2\alpha}\right),2\frac{\beta_c}{\beta_s}\right\}\left[E_s+\eta^2\right]. \tag{5.36}$$

The Ineq. (5.19) follows directly from Eq. (5.36).
□

From Ineqs. (5.11) and (5.19), we have $V(t)$ constant or decreasing $\forall t\in[0,\infty)$ and lower bounded by zero. It follows from Ineq. (5.11) that $E_s(t)$ and $\eta(t)$ are bounded $\forall t\in[0,\infty)$. Since $E_s(t)$ is bounded $\forall t\in[0,\infty)$, we know that the potential and kinetic energies defined in Eq. (5.9) are bounded. We now use Ineq. (3.5) to show that $u(x,t)$ is bounded $\forall t\in[0,\infty)$ and $\forall x\in[0,L]$. We assume $u_t(x,t)\in\mathcal{S}([0,L])$ so bounded kinetic energy means $u_t(x,t)\in\mathcal{L}_\infty([0,L])$. We assume $u(x,t)\in\mathcal{S}_{A_0}([0,L])$ so $u_x(L,t)$ and $u_{xx}(x,t)$ $\forall x\in[0,L]$ are bounded $\forall t\in[0,\infty)$. Since $u_x(L,t)$ and $\eta(t)$ are bounded $\forall t\in[0,\infty)$, we use Eq. (5.4) to show that $u_t(L,t)$ is bounded $\forall t\in[0,\infty)$. Given $u_{xx}(x,t)\in\mathcal{L}_\infty([0,L])$, the field Eq. (5.1) implies that $u_{tt}(x,t)$ is bounded $\forall x\in[0,L]$ and $\forall t\in[0,\infty)$ and because $\dot{\eta}\in\mathcal{L}_\infty$ we obtain $u_{xt}(L,t)\in\mathcal{L}_\infty$. Thus, the control input of Eq. (5.6) is bounded $\forall t\in[0,\infty)$.

Using Eqs. (5.19) and (5.11) we obtain an exponential bound for V as follows

$$V(t)\leq V(0)\exp\left(-\frac{\lambda_3}{\lambda_2}t\right) \tag{5.37}$$

In addition, we use Eq. (5.9), and Ineqs. (5.11) and (3.5) to formulate the following inequalities

$$\frac{\beta_s\mathcal{B}A}{2L}u^2(x,t)\leq\frac{\beta_s\mathcal{B}A}{2}\int_0^L u_x^2dx\leq E_s(t)\leq\frac{1}{\lambda_1}V(t)\quad\forall x\in[0,L]. \tag{5.38}$$

We now use Eq. (5.37), and the inequality-type bound developed in Ineq. (5.38) to obtain the strong exponential stability result.
□

5.1.2 Fourth Order Systems

In this section we develop an EMK boundary controller for the fourth order cantilevered beam model. The control objective is to ensure strong exponential stability of the beam displacement.

Boundary Control of a Cantilevered Beam

The EMK controller for the cantilevered beam model of Eqs. (2.46) through (2.49) is

$$f = mw_{xxxt}(L) - EIw_{xxx}(L) - k_s\eta \tag{5.39}$$

where k_s is a positive control gain and the auxiliary tracking signal is

$$\eta = w_t(L) - w_{xxx}(L). \tag{5.40}$$

After differentiating Eq. (5.40) with respect to time, multiplying the resulting expression by m, and substituting the right-hand side of Eq. (2.49) for $mw_{tt}(L)$, we have

$$m\dot{\eta} = -mw_{xxxt}(L) + EIw_{xxx}(L) + f. \tag{5.41}$$

After substituting Eq. (5.39) into Eq. (5.41), we obtain the following closed-loop boundary dynamics

$$m\dot{\eta} = -k_s\eta. \tag{5.42}$$

Enforcement of $\eta = 0$ ensures a damped boundary condition at $x = L$.

Theorem 40 *Given the field equation dynamics of Eq. (2.47) and the boundary conditions given by Eqs. (2.48) and (2.49), the EMK boundary controller given by Eq. (5.39) ensures that the beam displacement is strongly exponentially stabilized as follows*

$$|w(x,t)| \leq \sqrt{\frac{2\lambda_2 L^3}{\alpha\lambda_1 EI}\kappa_o \exp\left(-\frac{\lambda_3}{\lambda_2}t\right)} \qquad \forall x \in [0, L] \tag{5.43}$$

where α, λ_1, λ_2, λ_3, and κ_o are positive constants.

Proof: We begin with the Lyapunov functional

$$V(t) = E_b(t) + \frac{1}{2}m\eta^2(t) + E_c(t), \tag{5.44}$$

where

$$E_b = \frac{\alpha}{2}\int_0^L \rho w_t^2 dx + \frac{\alpha}{2}\int_0^L EIw_{xx}^2 dx, \tag{5.45}$$

and

$$E_c = 2\beta \int_0^L \rho x w_t w_x dx, \tag{5.46}$$

where α and β are positive constants. If we choose β sufficiently small, we can formulate the following upper and lower bound on $V(t)$ given in Eq. (5.44) for some positive constants λ_1 and λ_2.

Lemma 41 *The function given by Eq. (5.44) can be upper and lower bounded as follows*

$$\lambda_1 \left(E_b + \eta^2\right) \leq V(t) \leq \lambda_2 \left(E_b + \eta^2\right), \tag{5.47}$$

where λ_1 and λ_2 are positive constants.

Proof: First, note that Ineq. (3.10) can be used to bound $E_c(t)$ of (5.46) as follows

$$E_c = 2\beta \int_0^L \rho x w_t w_x dx \leq 2\beta\rho L \int_0^L \left(w_t^2 + w_x^2\right) dx. \tag{5.48}$$

After applying Ineq. (3.24) to the w_x^2 term on the right-hand side of Eq. (5.48), we have

$$\begin{aligned} E_c &\leq 2\beta\rho L \int_0^L \left(w_t^2 + L^2 w_{xx}^2\right) dx \\ &\leq 2\beta\rho L \max(1, L^2) \int_0^L \left(w_t^2 + w_{xx}^2\right) dx \\ &\leq 4\beta\rho L \frac{\max(1, L^2)}{\min(\rho, EI)} \frac{1}{2} \int_0^L \left(\rho w_t^2 + EI w_{xx}^2\right) dx. \end{aligned} \tag{5.49}$$

We can now use Eq. (5.45) and Eq. (5.49) to establish the following inequality

$$-\xi E_b \leq E_c \leq \xi E_b, \tag{5.50}$$

where $\xi = 4\alpha\beta\rho L \dfrac{\max(1, L^2)}{\min(\rho, EI)}$. Thus, for sufficiently small β, we have

$$0 \leq \zeta_1 E_b \leq E_b + E_c \leq \zeta_2 E_b \tag{5.51}$$

for some positive constants $\xi_1 = 1 - \xi$ and $\xi_2 = 1 + \xi$. Given the structure of $V(t)$ defined in Eq. (5.44) and the inequality given by Eq. (5.51), the inequality given by Eq. (5.47) is obvious.

□

Lemma 42 *The time derivative of the Lyapunov functional given by Eq. (5.44) can be upper bounded as follows*

$$\dot{V} \leq -\lambda_3 \left(E_b + \eta^2\right). \tag{5.52}$$

where $\lambda_3 > 0$.

Proof: After differentiating Eq. (5.44) with respect to time, we have

$$\dot{V} = \dot{E}_b + \dot{E}_c - k_s\eta^2. \tag{5.53}$$

where Eq. (5.42) has been utilized. To determine $\dot{E}_b(t)$ in Eq. (5.53), we differentiate Eq. (5.45) with respect to time to obtain

$$\dot{E}_b = -\alpha EI \int_0^L w_t w_{xxxx} dx + \alpha EI \int_0^L w_{xx} w_{xxt} dx \tag{5.54}$$

where Eq. (2.47) has been utilized. Integrating the first term on the right-hand side of Eq. (5.54) by parts twice, we obtain

$$\begin{aligned} \dot{E}_b &= -\alpha EI w_t(L) w_{xxx}(L) \\ &= -\frac{\alpha EI}{2}\left(w_t^2(L) + w_{xxx}^2(L)\right) + \frac{\alpha EI}{2}\eta^2 \end{aligned} \tag{5.55}$$

after applying the boundary conditions (2.48) and Eq. (5.40).

After differentiating Eq. (5.46) with respect to time, we obtain

$$\dot{E}_c = A_1 + A_2 \tag{5.56}$$

where

$$\begin{aligned} A_1 &= 2\beta\rho \int_0^L x w_t w_{xt} dx \\ A_2 &= -2\beta EI \int_0^L x w_x w_{xxxx} dx \end{aligned} \tag{5.57}$$

after using Eq. (2.47) in the expression for A_2. After integrating the expression for A_1 given by Eq. (5.57) by parts, we obtain

$$A_1 = 2\beta\rho \left(L w_t^2(L) - \int_0^L w_t^2 dx \right) - 2\beta\rho \int_0^L x w_{xt} w_t dx. \tag{5.58}$$

The last term in Eq. (5.58) equals A_1, so we can rewrite Eq. (5.58)

$$A_1 = \beta\rho\left(Lw_t^2(L) - \int_0^L w_t^2 dx\right). \tag{5.59}$$

Integrating Eq. (5.57) by parts, we obtain

$$\begin{aligned} A_2 &= -2\beta EILw_x(L)w_{xxx}(L) && (5.60) \\ &\quad +2\beta EI\left(\int_0^L w_x w_{xxx} dx + \int_0^L x w_{xx} w_{xxx} dx\right). && (5.61) \end{aligned}$$

Integration of the first integral in Eq. (5.60) by parts produces

$$\begin{aligned} A_2 &= -2\beta EILw_x(L)w_{xxx}(L) \\ &\quad +2\beta EI\left(-\int_0^L w_{xx}^2 dx + \int_0^L x w_{xx} w_{xxx} dx\right) \end{aligned} \tag{5.62}$$

The last integral on the right-hand side of Eq. (5.62) simplifies to

$$\begin{aligned} A_2 = &-2\beta EILw_x(L)w_{xxx}(L) \\ &-2\beta EI\left(2\int_0^L w_{xx}^2 dx + \int_0^L x w_{xx} w_{xxx} dx\right) \end{aligned} \tag{5.63}$$

Combining the expressions given by Eq. (5.62) and Eq. (5.63), we obtain

$$A_2 = -\beta EI\left(2Lw_x(L)w_{xxx}(L) + 3\int_0^L w_{xx}^2 dx\right) \tag{5.64}$$

Substituting Eqs. (5.55), (5.56), (5.59), and (5.64) into Eq. (5.53), we have

$$\begin{aligned} \dot{V} &= -\left(\frac{\alpha EI}{2} - \beta L\rho\right) w_t^2(L) - \frac{\alpha EI}{2} w_{xxx}^2(L) \\ &\quad -\left(k_s - \frac{\alpha EI}{2}\right)\eta^2 - \frac{\beta}{\alpha}E_b - 2\beta EI\int_0^L w_{xx}^2 dx \\ &\quad -2\beta EILw_x(L)w_{xxx}(L) \end{aligned} \tag{5.65}$$

where Eq. (5.45) has been used. Applying Ineq. (3.9) and Ineq. (3.12) to the second line of Eq. (5.65), produces the upper bound for $\dot{V}(t)$

$$\begin{aligned}\dot{V} \leq & -\left(\frac{\alpha EI}{2}-\beta L\rho\right)w_t^2(L)-\frac{\alpha EI}{2}w_{xxx}^2(L)\\ & -\left(k_s-\frac{\alpha EI}{2}\right)\eta^2-\frac{\beta}{\alpha}E_b-2\beta EI\frac{1}{L}w_x^2(L)\\ & +2EIL\beta\delta w_x^2(L)+\frac{2EIL\beta}{\delta}w_{xxx}^2(L)\end{aligned} \quad (5.66)$$

where δ is an arbitrary positive constant. Rearrangement of Eq. (5.66) results in

$$\begin{aligned}\dot{V} \leq & -\left(\frac{\alpha EI}{2}-\beta L\rho\right)w_t^2(L)-\left(k_s-\frac{\alpha EI}{2}\right)\eta^2\\ & -\frac{\beta}{\alpha}E_b-2EIL\beta\left(\frac{1}{L^2}-\delta\right)w_x^2(L)\\ & -EI\left(\frac{\alpha}{2}-\frac{2L\beta}{\delta}\right)w_{xxx}^2(L).\end{aligned} \quad (5.67)$$

From Eq. (5.67), if we chose the design constants $\delta < \frac{1}{L^2}$, $\alpha < \frac{2k_s}{EI}$, and $\beta < \min\left(\frac{\alpha EI}{2L\rho}, \frac{\alpha\delta}{4L}\right)$ then we obtain the upper bound given in Eq. (5.52).
$\square$

From Eq. (5.47) and Eq. (5.52), we can obtain the following upper bound for the time derivative of $V(t)$

$$\dot{V}(t) \leq -\frac{\lambda_3}{\lambda_2}V(t) \quad (5.68)$$

whose solution yields

$$\begin{aligned}V(t) & \leq V(0)\exp\left(-\frac{\lambda_3}{\lambda_2}t\right)\\ & \leq \lambda_2\left(E_b(0)+\eta^2(0)\right)\exp\left(-\frac{\lambda_3}{\lambda_2}t\right)\end{aligned} \quad (5.69)$$

where Eq. (5.47) allows formulation of the inequality on the right-hand side of Eq. (5.69). In addition, we can use Eqs. (3.5), (5.45), and (5.47) to generate

$$\frac{1}{2}\alpha EI\frac{1}{L^3}w^2(x,t) \leq \frac{\alpha}{2}\int_0^L EIw_{xx}^2dx \leq E_b(t) \leq \frac{1}{\lambda_1}V(t), \quad (5.70)$$

for all $x \in [0, L]$. Ineq. (5.43) follows from combination of Eqs. (5.69) and (5.70) and use of the explicit definitions given by Eq. (5.40) and Eq. (5.45).

□

We have shown that the EMK control law Eq. (5.39) exponentially stabilizes the distributed response. Now we show that $w_{xxxt}(L,t)$, $w_{xxx}(L,t)$, $w_t(L,t)$, and, hence, the control force are bounded. From Eq. (5.43) we know $w(x,t) \in \mathcal{L}_\infty([0,L])$ so $\mathbf{w}(\mathbf{x},t) = [w(x,t), w(L,t)]^T \in \mathcal{L}^2_\infty([0,L])$. If we assume $\mathbf{w} \in \mathcal{S}_{\mathcal{A}_0}([0,L])$ then

$$\mathcal{A}_0\mathbf{w} = \begin{bmatrix} \frac{EI}{\rho} w_{xxxx} \\ -\frac{EI}{m} w_{xxx}(L,t) \end{bmatrix} \in \mathcal{L}^2_\infty([0,L])$$

and $w_{xxx}(L,t) \in \mathcal{L}_\infty$. From Eqs. (5.69) and (5.44), we can state that $E_b(t)$ and $\eta(t)$ are bounded, implying the kinetic energy and $w_t(L,t) \in \mathcal{L}_\infty$, respectively. Boundedness of the kinetic energy means

$$\|\mathbf{w}_t\| = \frac{1}{2}\int_0^L \rho w_t^2 dx + \frac{1}{2} m w_t^2(L) \in \mathcal{L}_\infty.$$

If we assume that $\mathbf{w}_t \in \mathcal{S}^2([0,L])$, then $\mathbf{w}_t = [w_t(x,t), w_t(L,t)]^T \in \mathcal{L}^2_\infty([0,L])$. Finally, assuming $\mathbf{w}_t \in \mathcal{S}_{\mathcal{A}_0}([0,L])$, implies $w_{xxxt}(L,t) \in \mathcal{L}_\infty$.

5.1.3 Backstepping Compensation for Actuator Dynamics

In the previous work, we assume that the actuator responds instantaneously to torque or force commands. In practice, however, actuators such as electric motors and speakers have inductance that produces a time lag between the commanded and actual force. In many cases, the actuator electrical dynamics are significant and cannot be neglected. Distributed models have infinite bandwidth so even fast actuator dynamics interact with corresponding high frequency modes.

In this section, we develop a backstepping controller that compensates for actuator electromechanical dynamics in a boundary noise control system. Backstepping requires that we differentiate the control force input. Thus, we avoid $u_{xt}(L,t)$ terms in the control law because $u_{xtt}(L,t)$ terms arise. There are no dynamics for $u_x(L,t)$ that can relate it to lower derivative, more easily measurable signals. Without the $u_{xt}(L,t)$ term in the control law, however, we cannot obtain strong exponential stability using a Lyapunov approach. We therefore rely on LaSalle's invariance principle to prove asymptotic stability.

Acoustic Duct with Speaker Dynamics

The speaker electrical dynamics enter the system model through the boundary condition Eq. (5.3) as follows

$$mu_{tt}(L,t) + \mathcal{B}Au_x(L,t) + B_d u_t(L,t) + K_s u(L,t) = \kappa_\tau I(t), \tag{5.71}$$

where the force Eq. (2.6) is used. The loudspeaker electrical dynamics are given in Eq. (2.5).

Backstepping Control Development To facilitate the development of the backstepping controller, we first define the current tracking error $\nu(t)$ as follows

$$\nu(t) = I_d(t) - I(t), \tag{5.72}$$

where $I_d(t)$ denotes the desired loudspeaker current. After substituting Eq. (5.72) into the boundary dynamics given in Eq. (5.3), we obtain the following open loop dynamics for $u(L,t)$

$$\begin{array}{l} mu_{tt}(L,t) + \mathcal{B}Au_x(L,t) + B_d u_t(L,t) \\ \quad +K_s u(L,t) + \kappa_\tau \nu(t) = \kappa_\tau I_d(t). \end{array} \tag{5.73}$$

Based on the structure of Eq. (5.73), the control objective and the subsequent stability analysis, we can design the desired loudspeaker current $I_d(t)$ as follows

$$\begin{array}{rl} I_d(t) = & \frac{1}{\kappa_\tau}\left[B_d u_t(L,t) + K_s u(L,t)\right. \\ & \left.-k_v u_t(L,t) - k_p u_x(L,t)\right], \end{array} \tag{5.74}$$

where k_p and k_v are positive constant feedback gains. Implementation of the desired current control law of Eq. (5.74) requires the following boundary measurements: (i) $u(L,t)$, the loudspeaker baffle displacement, (ii) $u_t(L,t)$, the baffle velocity, and (iii) $u_x(L,t)$, the displacement gradient at $x = L$. Note that, to calculate $u_x(L,t)$, we use the standard acoustic pressure relationship $u_x(L,t) = P(t)/\mathcal{B}$, where $P(t)$ denotes the boundary pressure measured by a microphone. After substituting Eq. (5.74) into Eq. (5.73), we obtain the closed loop dynamics

$$mu_{tt}(L,t) + k_v u_t(L,t) + (\mathcal{B}A + k_p)\, u_x(L,t) + \kappa_\tau \nu(t) = 0$$

To facilitate the subsequent stability analysis, we define the functional

$$V_1(t) = \frac{\rho}{2}\int_0^L u_t^2(x,t)\,dx + \frac{\mathcal{B}}{2}\int_0^L u_x^2(x,t)\,dx + \frac{1}{2}m\frac{\mathcal{B}}{\mathcal{B}A+k_p}u_t^2(L,t). \tag{5.75}$$

Taking the time derivative of Eq. (5.75), substituting for $\rho u_{tt}(x,t)$ from Eq. (5.1), and for $mu_{tt}(L,t)$ from Eq. (5.73), we obtain

$$\begin{array}{rcl} \dot{V}_1(t) & = & \mathcal{B}u_x(L,t)\,u_t(L,t) + \frac{\mathcal{B}}{\mathcal{B}A+k_p}\left[-\mathcal{B}Au_x(L,t) - B_d u_t(L,t)\right. \\ & & \left.-K_s u(L,t) + \kappa_\tau I_d(t) - \kappa_\tau \nu(t)\right] u_t(L,t). \end{array} \tag{5.76}$$

Substitution of Eq. (5.74) into Eq. (5.76) produces

$$\dot{V}_1(t) = -\frac{\mathcal{B}}{\mathcal{B}A + k_p}\left[k_v u_t^2(L,t) + \kappa_\tau \nu(t)\, u_t(L,t)\right]. \tag{5.77}$$

To facilitate the design of the desired loudspeaker voltage, $E(t)$, we differentiate the current tracking error given in Eq. (5.72) and use Eq. (2.5) to obtain the open loop current tracking error dynamics

$$L_a\dot{\nu}(t) = L_a\dot{I}_d(t) + RI(t) + \kappa_\tau u_t(L,t) - E(t). \tag{5.78}$$

Based on the structure of Eq. (5.78), we design the desired voltage $E(t)$ as follows

$$\begin{aligned} E(t) &= L_a \dot{I}_d(t) + RI(t) + k_e \nu(t), \\ &= -\left(\frac{BAL_a(B_d - k_v)}{\kappa_\tau m}\right) u_x(L,t) - \left(\frac{L_a k_p}{\kappa_\tau}\right) u_{xt}(L,t) \\ &\quad - \left(\frac{B_d L_a(B_d - k_v)}{\kappa_\tau m} - \frac{L_a K_s}{\kappa_\tau}\right) u_t(L,t) + k_e \nu(t) \\ &\quad - \left(\frac{K_s L_a(B_d - k_v)}{\kappa_\tau m}\right) u(L,t) + \left(R + \frac{L_a(B_d - k_v)}{m}\right) I(t), \end{aligned} \tag{5.79}$$

where the back-emf constant $k_e = 1/k_\tau$. Implementation of the desired voltage control law of Eq. (5.79) requires the measurement of: (i) $I(t)$, the current through the loudspeaker coils that can be measured by, for example, hall-effect current sensors, (ii) $u(L,t)$, loudspeaker baffle displacement, (iii) $u_t(L,t)$, the baffle velocity, (iv) $u_x(L,t)$, and (iv) $u_{xt}(L,t)$ that are obtained from boundary pressure measurements. Substituting Eq. (5.79) into Eq. (5.78), we obtain the closed loop dynamics

$$L_a \dot{\nu}(t) = \kappa_\tau u_t(L,t) - k_e \nu(t). \tag{5.80}$$

We now define a Lyapunov functional

$$V(t) = V_1(t) + \frac{1}{2} L_a \frac{\mathcal{B}}{\mathcal{B}A + k_p} \nu^2(t), \tag{5.81}$$

Taking the time derivative of Eq. (5.81), substituting for $\dot{V}_1(t)$ from Eq. (5.77) and for $L_a \dot{\nu}(t)$ from Eq. (5.80), we obtain

$$\dot{V}(t) = -\frac{\mathcal{B}}{\mathcal{B}A + k_p}\left[k_v u_t^2(L,t) + k_e \nu^2(t)\right] \leq 0. \tag{5.82}$$

Thus, the backstepping control law ensures that the closed loop system is stable. The negative semi-definite result in Eq. (5.82), however, does not prove asymptotic stability. Using LaSalle's invariance principle, we can show that the closed loop acoustic displacement $u(x,t)$ is driven asymptotically to zero [93].

Simulation Results To simulate the proposed controller, Galerkin's method, discussed in Chapter 2, is used to obtain a finite-dimensional approximation for the mechanical subsystem dynamics of Eqs. (5.1) through (5.3). The solution of the field equation given in Eq. (5.1) is of the form

$$u(x,t) = U(x) \exp\left(i\sqrt{\tfrac{\mathcal{B}}{\rho}}\beta t\right), \tag{5.83}$$

where $U(x)$ is the open loop eigenfunction and β is the open loop eigenvalue of $u(x,t)$. Satisfaction of the boundary conditions of Eqs. (5.2) and (5.3) (with $f(t) = I(t) = B_d = K_s = 0$) yields the following form for the eigenfunctions

$$U_i(x) = \sin(\beta_i x), \tag{5.84}$$

where the eigenvalues satisfy the frequency equation

$$\rho A \cos(\beta L) - m\beta \sin(\beta L) = 0. \tag{5.85}$$

To obtain the finite dimensional, closed loop system dynamics, we first approximate the variable $u(x,t)$ with the following n-term expansion

$$u(x,t) = \textstyle\sum_{i=1}^{n} U_i(x)\, q_i(t), \tag{5.86}$$

where $q_i(t)$ denotes the generalized coordinates. Upon substitution of the control inputs given in (5.74), we obtain the following ordinary differential equation representing the finite dimensional mechanical subsystem dynamics

$$M_c\ddot{\mathbf{q}} + B_c\dot{\mathbf{q}} + K_c\mathbf{q} + N_c\nu = \mathbf{F}. \tag{5.87}$$

where

$$\mathbf{q}(t) = \left[\; q_1(t), \;\; \ldots.., \;\; q_n(t) \;\right]^T, \tag{5.88}$$

$$\mathbf{F}(t) = \left[\; U_1(x_d), \;\; \ldots.., \;\; U_n(x_d) \;\right]^T f(t), \tag{5.89}$$

$$[M_c]_{ij} = \rho \textstyle\int_0^L U_i(x)\, U_j(x)\, dx + mU_i(L)\, U_j(L), \tag{5.90}$$

$$[B_c]_{ij} = k_v U_i(L)\, U_j(L), \tag{5.91}$$

$$[K_c]_{ij} = \mathcal{B} \textstyle\int_0^L \beta_i^2 U_i(x)\, U_j(x)\, dx + (\mathcal{B}A + k_p)\, U_{ix}(L)\, U_j(L), \tag{5.92}$$

and the vector $N_c \in \Re^n$, which represents the coupling between the electrical and mechanical subsystems, is given by

$$[N_c]_j = \kappa_\tau U_j(L), \tag{5.93}$$

The closed loop electrical subsystem is simulated using the current tracking error dynamics given by Eq. (5.80)

$$L_a\dot{\nu} + k_e\nu - N_c^T\dot{\mathbf{q}} = 0. \tag{5.94}$$

The system parameters used in the simulations are: $m = 0.085$ [kg], $\rho = 1.21$ $\left[\text{kg/m}^3\right]$, $\mathcal{B} = 1.43 \times 10^5$ [Pa], $K_s = 1.51 \times 10^4$ [N/m], $L_a = 0.256$ [mH], $R = 4.0$ [Ω], $\kappa_\tau = 8.55$ [N/A], $L = 2$ [m], $A = 0.114$ [m^2], $B_d = 1.51$ [N/m-s].

The following simulation results illustrate the response of the system's first ten vibration modes ($n = 10$) to a simulated impulse at $x = 0.1L$. For comparison purposes, the system is first simulated in open loop ($E = 0$) and the natural damping in the loudspeaker baffle is the only contributing factor to noise decay. Plots (a), (b) and (c) of Figure 2 show the open loop loudspeaker baffle displacement, $u(L,t)$, and the pressure decay at midspan and the actuated end. Next, a standard damper control (*i.e.*, baffle velocity feedback) law given by

$$\begin{aligned} I_d(t) &= -k_d u_t(L,t), \\ E(t) &= k_e\nu(t) \end{aligned} \tag{5.95}$$

is implemented with the control gains $k_v = 5.0$ and $k_e = 10.0$ that give the fastest transient decay with the 45 [V] available amplifier voltage. Plots (d), (e) and (f) of Figure 2 show the damper control induced loudspeaker baffle displacement, $u(L,t)$, and the pressure decay at midspan and the actuated end. The transient decay improves by a factor of four over the open loop.

Finally, the backstepping control law given by Eqs. (5.74) and (5.79), is implemented with control gains $k_p = 1.25 \times 10^4$, $k_v = 5.0$ and $k_e = 0.05$. Plots (a), (b) and (c) of Figure 3 show the backstepping boundary control induced loudspeaker baffle displacement, $u(L,t)$, and the pressure decay at midspan and at the actuated end. Plots (d) and (e) of Figure 3 show the loudspeaker voltage and the desired and actual loudspeaker currents in the closed loop.

The backstepping boundary control law drives the noise level to zero in approximately 0.1 seconds using less than 45 [V] and 430 [watts] peak. The backstepping boundary control transient decays two orders of magnitude faster than the open loop and twenty five times faster than the damper control.

The system performance in the frequency domain is shown in Figure 4. The loudspeaker pressure frequency response to the input disturbance is defined by the following transfer function

$$H_p(s) = \frac{P(L,s)}{F(s)}, \tag{5.96}$$

where $P(L,s)$ and $F(s)$ are the Laplace transforms of $\mathcal{B}u_x(L,t)$, the pressure at the speaker $(x = L)$ and $F(t)$, respectively. The frequency response plot shows that the backstepping boundary controller significantly attenuates the response peaks relative to open loop and damper control.

5.1.4 Compensation for Geometric and Material Nonlinearities

One of the advantages of using a Lyapunov approach in the formulation of control systems is that it can be applied to nonlinear systems. In this section we modify the boundary controlled string system governed by Eqs. (2.14) - (2.16) to include geometric and material nonlinearities and develop an EMK controller that strongly asymptotically stabilizes the string response.

Boundary Control of a Nonlinear String

System Model First we neglect tangential string motion [96] and assume that the tension is small relative to the axial stiffness as follows:

$$P(t) << EA, \tag{5.97}$$

so that the tension is uniform (*i.e.*, independent of x). We also assume that the tension in the string is related to the string stretch given by

$$y(t) = \int_0^L \left(\sqrt{1 + w_x^2} - 1\right) dx > 0, \tag{5.98}$$

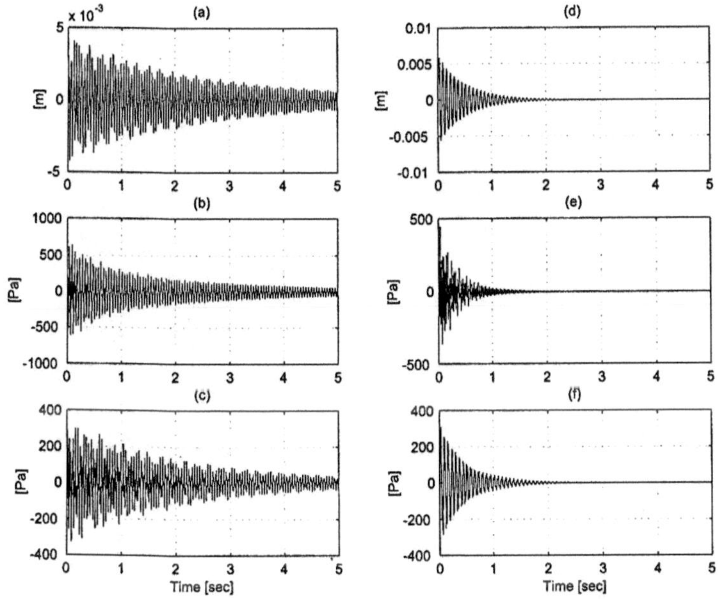

Figure 2 (a) Open loop loudspeaker baffle displacement, (b) open loop pressure measured at $x = \frac{L}{2}$, (c) open loop pressure measured at $x = L$, (d) damper control loudspeaker baffle displacement, (e) damper control pressure measured at $x = \frac{L}{2}$, (f) damper control pressure measured at $x = L$.

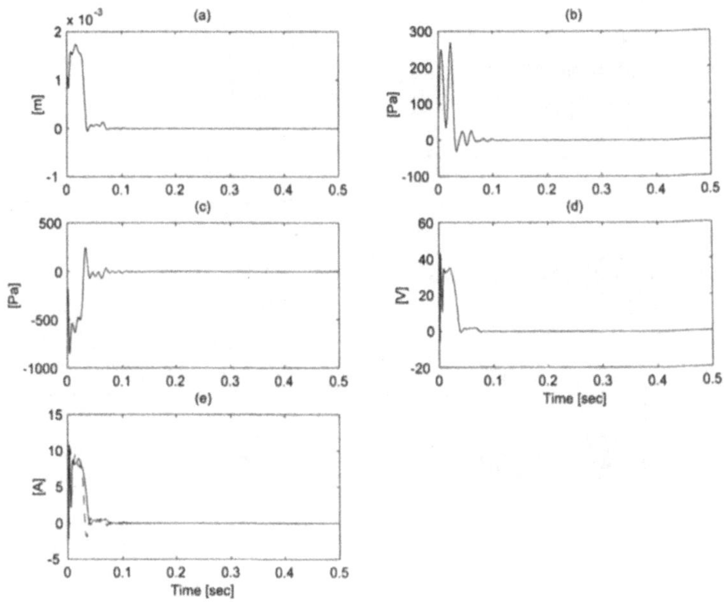

Figure 3 Backstepping boundary control: (a) loudspeaker baffle displacement, (b) pressure measured at $x = \frac{L}{2}$, (c) pressure measured at $x = L$, (d) applied loudspeaker voltage, and (e) desired (dashed line) and actual (solid line) loudspeaker currents.

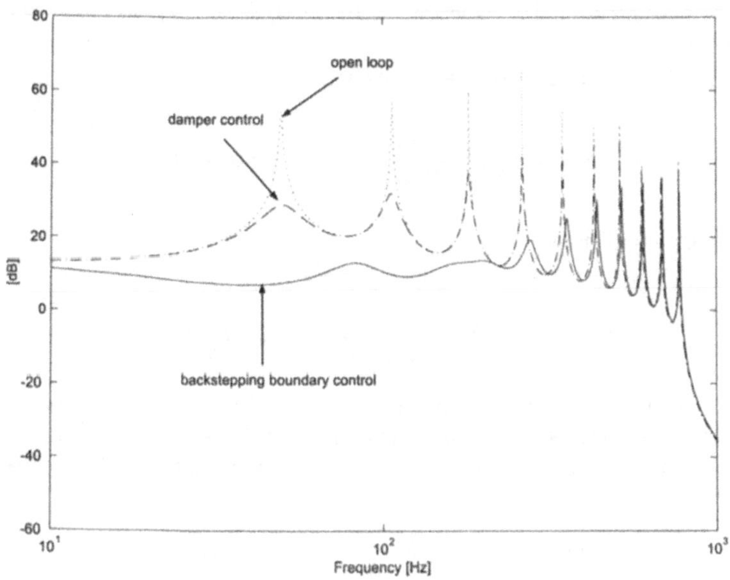

Figure 4 Frequency response of $H_p(j\omega)$: (a) open loop (dotted line), (b) damper control (dashed line), and (c) backstepping boundary control (solid line).

as follows

$$P(y(t)) \geq P_0 > 0 \tag{5.99}$$

where $y(t)$ is the string stretch. Finally, we assume that if $y(t)$ is bounded then $P(y(t))$ is also bounded and the potential energy stored can be lower and upper bounded as follows

$$\alpha_l y(t) \leq \Pi_p = \int_0^y P(y(t))\, dy \leq \kappa_u\left(y(t)\right) \tag{5.100}$$

where α_l is a positive scalar constant, $\kappa_u(0) = 0$, $\kappa_u(p) > 0\ \forall p > 0$, and $\kappa_u(p)$ is nondecreasing.

As in [19], the transverse field equation is

$$\rho w_{tt} = P(\cdot)\left[\frac{w_x}{\sqrt{1+w_x^2}}\right]_x \qquad \text{for } x \in (0, L), \tag{5.101}$$

with boundary conditions

$$w(0,t) = 0, \tag{5.102}$$

$$m w_{tt}(L) + P(\cdot)\frac{w_x(L)}{\sqrt{1+w_x^2(L)}} + Y(w_t(L))\phi = f(t), \tag{5.103}$$

where we have included the nondimensionalized actuator friction force $Y(w_t(L))\phi$ linearly parameterized as a known regression matrix $Y(\cdot) \in \Re^{1\times q}$ multiplied by a constant parameter vector $\phi \in \Re^{q\times 1}$. For example, an actuator with viscous damping b and Coulomb friction μ has

$$Y(\cdot) = \begin{bmatrix} w_t(L,t) & \operatorname{sgn}(w_t(L,t)) \end{bmatrix}, \tag{5.104}$$

$$\phi^T = \begin{bmatrix} b & \mu \end{bmatrix}, \tag{5.105}$$

where $\operatorname{sgn}(\cdot)$ is the signum function. We will also assume that $Y(\cdot)$ remains bounded $\forall t \in [0,\infty)$ if $w_t(L,t)$ is bounded $\forall t \in [0,\infty)$.

It can be shown that the dynamics given by Eqs. (5.101), (5.102), and (5.103) reduce to the linear string model if a linear stress-strain relationship is used:

$$P(y(t)) = P_0 + \frac{EA}{L}y(t), \tag{5.106}$$

and the slope of the transverse displacement is small (*i.e.*, $w_x << 1$ means that $\sqrt{1+w_x^2(L)} \approx 1$ in Eqs. (5.101), (5.103), and (5.98)). The $P(y(t))$ of Eq. (5.106) is strictly positive and satisfies Eq. (5.100) for $\alpha_l = EA/L$ and $\kappa_u = EAy/L$. Most stress-strain models (*e.g.*, polynomial) satisfy the mild assumptions imposed by Eqs. (5.99) and (5.100).

Control Formulation The primary control objective is to design an asymptotically stabilizing controller for the dynamics of Eqs. (5.101), (5.102), and (5.103). Based on the subsequent stability analysis, we propose the following nonlinear, boundary controller

$$f(t) = -\frac{mw_{xt}(L)}{(1+w_x^2(L))^{3/2}} + P(\cdot)\frac{w_x(L)}{\sqrt{1+w_x^2(L)}} + Y(\cdot)\phi - \left(k + \frac{k_r}{2}P(\cdot)\right)\eta(t) \quad (5.107)$$

where

$$\eta(t) = w_t(L) + \frac{w_x(L)}{\sqrt{1+w_x^2(L)}}, \quad (5.108)$$

and k and k_r are scalar control gains chosen to satisfy

$$k > 0 \quad \text{and} \quad k_r \geq 1. \quad (5.109)$$

In the control law (5.107) we require one more sensor than the linear case to measure $P(t)$.

After differentiation of Eq. (5.108) with respect to time, multiplication of both sides by m, and substitution for $mw_{tt}(L)$ from the boundary condition Eq. (5.103), we obtain the open-loop dynamics

$$m\dot{\eta}(t) = \frac{mw_{xt}(L)}{(1+w_x^2(L))^{3/2}} - P(\cdot)\frac{w_x(L)}{\sqrt{1+w_x^2(L)}} - Y(\cdot)\phi + f(t) \quad (5.110)$$

Substitution of the control input (5.107) into the right-hand side of Eq. (5.110), produces the closed-loop dynamics

$$m\dot{\eta}(t) = -\left(k + \frac{k_r}{2}P(\cdot)\right)\eta(t). \quad (5.111)$$

Theorem 43 *Given the field Eq. (5.101), boundary condition Eq. (5.102), and closed-loop boundary condition Eq. (5.111), the kinetic energy Eq. (2.7) and the potential energy Eq. (5.100) decay asymptotically to zero.*

Proof: We first prove that all signals in the closed-loop system are bounded. We define the following non-negative, scalar function

$$V_1(t) = E(t) + \frac{1}{2}m\eta^2(t) \quad (5.112)$$

where

$$E(t) = \frac{1}{2}\int_0^L \rho w_t^2 dx + \int_0^y P(\cdot)dy. \quad (5.113)$$

From the structure of $V_1(t)$ given by Eqs. (5.112) and (5.113), we can see that $V_1(t) \geq 0$.

Lemma 44 *The time derivative of the function $V_1(t)$ defined in Eq. (5.112) can be upper bounded as $\dot{V}_1(t) \leq 0$.*

Proof: After differentiating Eq. (5.112) with respect to time, we obtain

$$\dot{V}_1 = \dot{E} - \left(k + \frac{k_r}{2}P(\cdot)\right)\eta^2 \tag{5.114}$$

where Eq. (5.111) has been utilized. To obtain the expression for $\dot{E}(t)$ in Eq. (5.114), we differentiate Eq. (5.113) with respect to time, and then substitute the field dynamics given by Eq. (5.101) and utilize the structure of $y(t)$ given in Eq. (5.98) to obtain

$$\dot{E} = P(\cdot)\left[\int_0^L w_t \frac{\partial}{\partial x}\left(\frac{w_x}{\sqrt{1+w_x^2}}\right)dx\right] + P(\cdot)\int_0^L \frac{w_{xt}w_x}{\sqrt{1+w_x^2}}dx. \tag{5.115}$$

After integrating the bracketed term in Eq. (5.115) by parts and cancelling common terms, we obtain

$$\dot{E} = P(\cdot)w_t(L)\frac{w_x(L)}{\sqrt{1+w_x^2(L)}} \tag{5.116}$$

where the boundary condition given by Eq. (5.102) has been employed. We now rewrite Eq. (5.116) as follows

$$\dot{E} = \frac{P(\cdot)}{2}\eta^2 - \frac{P(\cdot)}{2}\left(w_t^2(L) + \frac{w_x^2(L)}{1+w_x^2(L)}\right). \tag{5.117}$$

Substitution of Eq. (5.117) into Eq. (5.114) produces

$$\dot{V}_1 = -\left(k + \frac{(k_r - 1)}{2}P(\cdot)\right)\eta^2 - \frac{P(\cdot)}{2}\left(w_t^2(L) + \frac{w_x^2(L)}{1+w_x^2(L)}\right) \leq 0 \tag{5.118}$$

because k and k_r have been restricted according to Eq. (5.109) and $P(\cdot)$ is positive according to Eq. (5.99).
□

From Eq. (5.118) we know that $\dot{V}_1(t) \leq 0$ so $V_1(t) \in \mathcal{L}_\infty$. From Eq. (5.112), $E(t)$ of Eq. (5.113) and $\eta(t)$ of Eq. (5.108) are also bounded for all time. Since $E(t) \in \mathcal{L}_\infty$, it follows that the potential energy of the system defined in Eq. (5.100) is bounded. The potential energy is not an explicit function of $w(x,t)$ so we cannot invoke the standard assumptions used to prove signal boundedness. We therefore assume that boundedness of the potential energy implies that $w(x,t)$, $w_x(x,t)$, and $w_{xx}(x,t) \in \mathcal{L}_\infty([0,L])$. From the boundedness of $\eta(t)$, we use Eq. (5.108) to show that $w_t(L,t)$ is bounded. Since $E(t), \eta(t) \in \mathcal{L}_\infty$, the kinetic energy of the system given by Eq. (2.7) is bounded. Again, we assume that boundedness of the kinetic energy implies $w_t(x,t)$ and $w_{xt}(x,t) \in \mathcal{L}_\infty([0,L])$. With $E \in \mathcal{L}_\infty$, $y \in \mathcal{L}_\infty$ from Eq. (5.100) and $P(\cdot) \in \mathcal{L}_\infty$ from Eq. (5.99). The control force $f(t)$ given by Eq. (5.107) is bounded because $w_x(L,t)$, $w_t(L,t)$, $w_{xt}(L,t)$, $\eta(t)$, and $P(\cdot) \in \mathcal{L}_\infty$. Finally, since $\eta(t)$ and $P(\cdot) \in \mathcal{L}_\infty$, we can use Eq. (5.111) to state that $\dot{\eta}(t) \in \mathcal{L}_\infty$. We can now use Eq. (5.101) and Eq. (5.103) to show that $w_{tt}(x,t) \in \mathcal{L}_\infty([0,L])$.

To prove that the kinetic energy and the potential energy of the string-actuator system decay asymptotically to zero, we define another scalar functional

$$V(t) = V_1(t) + E_c(t) \tag{5.119}$$

where we use the crossing term

$$E_c(t) = 2\beta\rho \int_0^L x w_t w_x dx \tag{5.120}$$

and β is a positive weighting constant.

Lemma 45 *For sufficiently small β, the functional $V(t)$ defined in Eq. (5.119) can be bounded as given by*

$$0 \le \lambda_1(E_a(t) + \eta^2(t)) \le V(t) \le \lambda_2(E(t) + \eta^2(t)), \tag{5.121}$$

where

$$E_a(t) = \frac{1}{2}\int_0^L \rho w_t^2 dx + y(t) \tag{5.122}$$

and λ_1 and λ_2 are positive constants.

Proof: We first show that the summation of $E(t)$ and $E_c(t)$, defined in Eq. (5.113) and (5.120), respectively, can be bounded by

$$0 \le \mu_1 E_a \le E + E_c \le 2\mu_2 E \tag{5.123}$$

where

$$\mu_1 = \min\left\{\left(\frac{1}{2} - \beta L\right), (\alpha_l - 2\beta\rho L\delta)\right\} > 0, \tag{5.124}$$

$$\mu_2 = \max\left\{\left(\frac{1}{2} + \beta L\right), \left(1 + 2\frac{\beta\rho L\delta}{\alpha_l}\right)\right\} > 0, \tag{5.125}$$

if β sufficiently small. To prove the inequality given by Eq. (5.123), we first note from Eq. (5.120) that

$$E_c \le 2\beta\rho \int_0^L x\left|w_t w_x\right| dx \le \beta\rho L\left(\int_0^L w_t^2 dx + \int_0^L w_x^2 dx\right). \tag{5.126}$$

We observe that from Theorem 43 that $w_x(x,t) \in \mathcal{L}_\infty([0,L])$ so

$$\delta = 1 + \sup_{x,t}\left\{w_x^2\right\} \tag{5.127}$$

is bounded. Use of Eq. (5.127) in Eqs. (5.126) and (5.128) produces

$$E_c \leq \beta\rho L \int_0^L w_t^2 dx + \beta\rho L\delta \int_0^L \left(\frac{w_x}{\sqrt{1+w_x^2}}\right)^2 dx \tag{5.128}$$

and

$$-\beta\rho L \int_0^L w_t^2 dx - \beta\rho L\delta \int_0^L \left(\frac{w_x}{\sqrt{1+w_x^2}}\right)^2 dx \leq E_c, \tag{5.129}$$

respectively. We now use Eq. (5.129) and Eq. (5.113) to formulate the following lower bound

$$\left(\frac{1}{2} - \beta L\right) \int_0^L \rho w_t^2 dx + \left[\int_0^y P(\cdot)dy - \beta\rho L\delta \int_0^L \left(\frac{w_x}{\sqrt{1+w_x^2}}\right)^2 dx\right] \leq E_c + E. \tag{5.130}$$

The bracketed term in Eq. (5.130) can be lower bounded by observing that

$$\left(\frac{w_x}{\sqrt{1+w_x^2}}\right)^2 \leq 2\left(\sqrt{1+w_x^2} - 1\right). \tag{5.131}$$

Multiplication of Eq. (5.131) by $-\beta\rho L\delta$ and adding $\alpha_l\left(\sqrt{1+w_x^2}-1\right)$ produces

$$(\alpha_l - 2\beta\rho L\delta)\left(\sqrt{1+w_x^2} - 1\right) \leq -\beta\rho L\delta\frac{w_x^2}{1+w_x^2} + \alpha_l\left(\sqrt{1+w_x^2} - 1\right). \tag{5.132}$$

Thus, if we restrict the weighting constant β according to

$$\beta < \frac{\alpha_l}{2\rho L\delta}, \tag{5.133}$$

then the left hand side of Eq. (5.132) will always be non-negative. Integration of Eq. (5.132) produces

$$0 \leq (\alpha_l - 2\beta\rho L\delta)\, y \leq -\beta\rho L\delta \int_0^L \left(\frac{w_x}{\sqrt{1+w_x^2}}\right)^2 dx + \alpha_l y. \tag{5.134}$$

Use of Eq. (5.100) on the right hand side of Eq. (5.134) produces

$$0 \leq (\alpha_l - 2\beta\rho L\delta)\, y \leq -\beta\rho L\delta \int_0^L \left(\frac{w_x}{\sqrt{1+w_x^2}}\right)^2 dx + \int_0^y P dy. \tag{5.135}$$

Substitution of Eq. (5.135) simplifies Eq. (5.130) to

$$0 \leq \left[\left(\frac{1}{2} - \beta L\right) \int_0^L \rho w_t^2 dx\right] + (\alpha_l - 2\beta\rho L\delta)\, y \leq E_c + E. \tag{5.136}$$

where we further restrict the weighting constant β according to

$$\beta < \frac{1}{2L}. \tag{5.137}$$

We can now use Eq. (5.136) to formulate the following lower bound

$$\mu_1 E_a \leq E + E_c. \tag{5.138}$$

To determine the upper bound given in Eq. (5.123), we use Eq. (5.128) and Eq. (5.113) to obtain

$$E + E_c \leq \left(\frac{1}{2} + \beta L\right) \int_0^L \rho w_t^2 dx + \int_0^y P dy + \beta\rho L\delta \int_0^L \left(\frac{w_x}{\sqrt{1+w_x^2}}\right)^2 dx. \tag{5.139}$$

Use of Eqs. (5.131), (5.100), (5.98), and (5.139) produces

$$\begin{aligned} E + E_c &\leq \left(\frac{1}{2} + \beta L\right) \int_0^L \rho w_t^2 dx + \int_0^y P dy + 2\beta\rho L\delta \int_0^L \left(\sqrt{1+w_x^2} - 1\right) dx \\ &\leq \left(\frac{1}{2} + \beta L\right) \int_0^L \rho w_t^2 dx + \left(1 + \frac{2\beta\rho L\delta}{\alpha_l}\right) \int_0^y P dy \\ &\leq \mu_2 \left(\int_0^L \rho w_t^2 dx + \int_0^y P dy\right) \leq 2\mu_2 E. \end{aligned} \tag{5.140}$$

Combination of Eqs. (5.138), (5.124), and (5.140), gives the result in Eq. (5.121).
□

Lemma 46 *For sufficiently small β, the time derivative of the functional $V(t)$ of Eq. (5.119) can be upper bounded as follows*

$$\dot{V}(t) \leq -\kappa_o \left(\int_0^L \left(\sqrt{1+w_x^2} - 1 \right) dx + \int_0^L w_t^2 dx + w_t^2(L) \right) \triangleq -g(t), \tag{5.141}$$

where κ_o a positive constant.

Proof: Differentiation of Eq. (5.119) with respect to time, we obtain

$$\dot{V} = \dot{V}_1 + \dot{E}_c. \tag{5.142}$$

From Theorem 43, we know that $\dot{V}_1(t)$ can be expressed as in Eq. (5.118). To obtain the expression for $\dot{E}_c(t)$ in Eq. (5.142), we differentiate Eq. (5.120) respect to time as follows

$$\dot{E}_c(t) = A_1 + A_2, \tag{5.143}$$

where

$$A_1 = 2\beta P \int_0^L x w_x \frac{\partial}{\partial x} \left(\frac{w_x}{\sqrt{1+w_x^2}} \right) dx, \qquad A_2 = \beta\rho \int_0^L x \frac{\partial}{\partial x} \left(w_t^2 \right) dx, \tag{5.144}$$

and Eq. (5.101) has been used. After integration of Eq. (5.144) by parts, we obtain

$$A_1 = 2\beta P L \frac{w_x^2(L)}{\sqrt{1+w_x^2(L)}} - 2\beta P \int_0^L \frac{w_x^2}{\sqrt{1+w_x^2}} dx - 2\beta P \int_0^L x \frac{\partial}{\partial x} \left(\sqrt{1+w_x^2} \right) dx \tag{5.145}$$

using the boundary conditions (5.102). Integrating the last term in Eq. (5.145) by parts, we obtain

$$\begin{aligned} A_1 &= 2\beta P L \frac{w_x^2(L)}{\sqrt{1+w_x^2(L)}} - 2\beta P L \sqrt{1+w_x^2(L)} \\ &\quad -2\beta P \int_0^L \left(\frac{w_x^2}{\sqrt{1+w_x^2}} - \sqrt{1+w_x^2} \right) dx + \left[2\beta P L - 2\beta P \int_0^L 1 dx \right] \end{aligned} \tag{5.146}$$

where the bracketed term (equal to zero) has been added to facilitate the analysis. Integration by parts Eq. (5.144) produces

$$A_2 = \beta\rho L \left(w_t(L) \right)^2 - \beta\rho \int_0^L \left(w_t \right)^2 dx. \tag{5.147}$$

We combine Eqs. (5.143), (5.146), and (5.147) to form

$$\dot{E}_c = 2\beta PL\frac{w_x^2(L)}{\sqrt{1+w_x^2(L)}} - 2\beta PL\left(\sqrt{1+w_x^2(L)}-1\right) + \beta\rho L w_t^2(L)$$
$$-\beta\rho\int_0^L w_t^2 dx - 2\beta P\int_0^L\left(\frac{w_x^2}{\sqrt{1+w_x^2}} + 1 - \sqrt{1+w_x^2}\right)dx. \quad (5.148)$$

After substituting Eqs. (5.148) and (5.118) into (5.142), we obtain

$$\dot{V} = -\left(k + \frac{k_r - 1}{2}P\right)\eta^2 - 2\beta PL\left(\sqrt{1+w_x^2(L)}-1\right) - \beta\rho\int_0^L w_t^2 dx$$
$$-\left(\frac{P}{2} - \beta\rho L\right)w_t^2(L) - \frac{P}{2}\left(\frac{1}{\sqrt{1+w_x^2(L)}} - 4\beta L\right)\frac{w_x^2(L)}{\sqrt{1+w_x^2(L)}}$$
$$-2\beta P\int_0^L\left(\frac{w_x^2}{\sqrt{1+w_x^2}} + 1 - \sqrt{1+w_x^2}\right)dx. \quad (5.149)$$

Application of Eqs. (5.99), (5.109), (5.127) produces

$$\dot{V} \leq -\kappa w_t^2(L) - \beta\rho\int_0^L w_t^2 dx - 2\beta P_o\int_0^L\left(\frac{w_x^2}{\sqrt{1+w_x^2}} + 1 - \sqrt{1+w_x^2}\right)dx, \quad (5.150)$$

where κ is some positive constant. Substitution of

$$\frac{w_x^2}{\sqrt{1+w_x^2}} + 1 - \sqrt{1+w_x^2} \geq \frac{1}{2}\left(\frac{\sqrt{1+w_x^2}-1}{\sqrt{1+w_x^2}}\right) \quad (5.151)$$

into Eq. (5.150) and using Eq. (5.127), we formulate the upper bound for $\dot{V}(t)$ as given in Eq. (5.141).

□

From $w_x(x,t)$, $w_t(x,t)$, $w_{xt}(x,t)$, $w_{tt}(x,t)$, and $w_{xx}(x,t) \in \mathcal{L}_\infty([0,L])$ we obtain $\dot{g}(t) \in \mathcal{L}_\infty$. Hence, from Eq. (5.141), we invoke Barbalat's Lemma to show that

$$\lim_{t\to\infty} g(t) = 0. \quad (5.152)$$

From the structure of $g(t)$ defined in Eq. (5.141), we use Eqs. (5.152), (5.113), (5.100) and (5.98) to show that the kinetic energy and the potential energy of the system decay asymptotically to zero.

□

Unfortunately, it is not clear how to strengthen this weakly asymptotically stable result. It seems reasonable that if the total energy is driven to zero, then the transverse displacement should also approach zero. There does not seem to be a straightforward way to use the result of Theorem 43, however, to prove $\lim_{t\to\infty}|w(x,t)| = 0$.

5.1.5 Effects of Nonuniform Parameter Distributions

Nonuniform system parameters occur frequently in practice. In the gantry crane, for example, the tension varies from a maximum at the gantry to a minimum at the payload. Beams are often designed with varying cross sections (e.g. tapered) to optimize structural properties. Unfortunately, however, nonuniformity greatly complicates the Lyapunov analysis. The challenge lies in determining an appropriate (nonuniform) cross term in the Lyapunov functional.

In this example, we allow the boundary controlled string to sag under gravity, making the tension vary along the length. An exponentially stabilizing EMK boundary controller for the model given in Eqs. (2.168) - (2.170) is derived.

Boundary Control of a Cable

The EMK control law is defined to be

$$f = -mw_{st}(1) + p(1)w_s(1) - k_s\left(w_t(1) + w_s(1)\right) + Y(w_t(1))\phi, \tag{5.153}$$

where k_s is a positive control gain and $Y(w_t(1))\phi$ compensates for nonlinear boundary friction as in the nonlinear string example. Note that the actuator mass m and tension $p(1)$ must be known to implement this controller.

Theorem 47 *Given the field Eq. (2.168) and the boundary conditions (2.169) and (2.170), the boundary controller (5.153) ensures that the cable displacement is strongly exponentially stabilized as follows*

$$|w(s,t)| \le \sqrt{\frac{2\lambda_2}{\lambda_1 p}\kappa_o \exp\left(-\frac{\lambda_3}{\lambda_2}t\right)} \qquad \forall s \in [0,1], \tag{5.154}$$

where λ_1, λ_2, λ_3, and κ_o are positive constants if the controller gain

$$k_s > \frac{p(1)}{2}. \tag{5.155}$$

Proof: To facilitate the stability proof, we first define the standard auxiliary tracking signal

$$\eta(t) = w_t(1) + w_s(1). \tag{5.156}$$

After differentiating Eq. (5.156) with respect to time, multiplying the resulting expression by m, and then substituting the right-hand side of Eq. (2.170) for $mw_{tt}(1)$, we have

$$m\dot{\eta} = mw_{st}(1) - p(1)w_s(1) - Y(w_t(1))\phi + f. \tag{5.157}$$

Substitution of Eq. (5.153) into Eq. (5.157) produces

$$m\dot{\eta} = -k_s\eta. \tag{5.158}$$

To prove Eq. (5.154), we begin with the Lyapunov functional

$$V(t) = E_s(t) + \frac{1}{2}m\eta^2(t) + E_c(t). \tag{5.159}$$

The cable vibration energy

$$E_s(t) = \frac{1}{2}\int_0^1 w_t^2 ds + \frac{1}{2}\int_0^1 p(s)w_s^2 ds, \tag{5.160}$$

and the crossing term

$$E_c(t) = 2\beta\int_0^1 \gamma(s)w_t w_s ds, \tag{5.161}$$

where β is a positive design constant and the weighting function

$$\begin{aligned} \gamma(s) &= \frac{h(s)}{2p_0}\left[\arctan\left(\frac{2s-1}{2p_0}\right) + p_0 \ln\left(2s - 1 + h(s)\right)\right. \\ &\quad \left. + \arctan\frac{1}{2p_0} + p_0 \ln\left(\sqrt{4p_0^2+1} - 1\right)\right] \end{aligned} \tag{5.162}$$

with

$$h(s) = \sqrt{4p_0^2 + 4s^2 - 4s + 1}.$$

The challenge of Lyapunov functional design for nonuniform systems is the development of an appropriate kernel $\gamma(s)$ for the crossing term functional. The difficulties associated with this iterative process are evident from the complexity of $\gamma(s)$ in Eq. (5.162).

Lemma 48 *For sufficiently small β, the nonnegative functional in Eq. (5.159) can be upper and lower bounded as given by*

$$\lambda_1\left(E_s(t) + \eta^2(t)\right) \le V(t) \le \lambda_2\left(E_s(t) + \eta^2(t)\right), \tag{5.163}$$

where λ_1 and λ_2 are positive constants.

Proof: First, note that Eq. (3.10) can be used to bound $E_c(t)$ of Eq. (5.161) as follows

$$\begin{aligned} E_c &= 2\beta \int_0^1 \gamma(s) w_t w_s ds \leq 2\beta\gamma(1) \int_0^1 \left(w_t^2 + w_s^2\right) ds, \\ &\leq \frac{4\beta\gamma(1)}{\min(1, p_0)} \left[\frac{1}{2}\int_0^1 \left(w_t^2 + p(s) w_s^2\right) ds\right], \end{aligned} \tag{5.164}$$

where we have used the facts that $p(s)$ is minimum at $s = 1/2$ for $s \in [0, 1]$ and that $\gamma(s)$ of Eq. (5.162) is maximum at $s = 1$ for $s \in [0, 1]$. We can now use Eq. (5.160) and Eq. (5.164) to establish the following inequality

$$-\frac{4\beta\gamma(1)}{\min(1, p_0)} E_s \leq E_c \leq \frac{4\beta\gamma(1)}{\min(1, p_0)} E_s. \tag{5.165}$$

If $\beta < \min\{1, p_0\} / 4\gamma(1)$ we have

$$\lambda_1 = 1 - \frac{4\beta\gamma(1)}{\min\{1, p_0\}} > 0 \text{ and } \lambda_2 = 1 + \frac{4\beta\gamma(1)}{\min\{1, p_0\}} \tag{5.166}$$

in Eq. (5.163).
□

Lemma 49 *The time derivative of the nonnegative function given by Eq. (5.159) can be upper bounded as given by*

$$\dot{V}(t) \leq -\lambda_3 \left(E_s(t) + \eta^2(t)\right), \tag{5.167}$$

where the positive constant λ_3 is positive if Ineq. (5.155) is satisfied.

Proof: After differentiating Eq. (5.159) with respect to time, we have

$$\dot{V} = \dot{E}_s + \dot{E}_c - k_s \eta^2, \tag{5.168}$$

where Eq. (5.158) has been used. Differentiation of Eq. (5.160) produces

$$\dot{E}_s = \int_0^1 w_t \left[p(s) w_s\right]_s ds + \int_0^1 p(s) w_s w_{st} ds, \tag{5.169}$$

where Eq. (2.168) has been used. If we integrate the first term on the right-hand side of Eq. (5.169) by parts, we obtain

$$\dot{E}_s = p(1) w_t(1) w_s(1) - p(0) w_t(0, t) w_s(0, t). \tag{5.170}$$

Application of the boundary conditions given in Eq. (2.169) to Eq. (5.170) produces

$$\dot{E}_s = p(1)w_t(1)w_s(1), \tag{5.171}$$

which can be written as

$$\dot{E}_s = -\frac{p(1)}{2}(w_t^2(1) + w_s^2(1)) + \frac{p(1)}{2}\eta^2 \tag{5.172}$$

using Eq. (5.156).

To determine $\dot{E}_c(t)$ in Eq. (5.168), we differentiate Eq. (5.161) with respect to time as follows

$$\dot{E}_c = A_1 + A_2, \tag{5.173}$$

where

$$A_1 = 2\beta \int_0^1 \gamma(s) w_t w_{st} ds, \qquad A_2 = 2\beta \int_0^1 \gamma(s) w_s \left[p(s) w_s\right]_s ds \ , \tag{5.174}$$

after using Eq. (2.168) in the expression for A_2. After integrating the expression for A_1 given by Eq. (5.174) by parts, we obtain

$$A_1 = 2\beta \left(\gamma(1) w_t^2(1) - \int_0^1 \gamma_s(s) w_t^2 ds \right) - 2\beta \int_0^1 \gamma(s) w_{st} w_t ds, \tag{5.175}$$

where we have used the fact that $\gamma(0) = 0$. After noting that the last term in Eq. (5.175) is equal to A_1, we can write Eq. (5.175) as follows

$$A_1 = \beta \left(\gamma(1) w_t^2(1) - \int_0^1 \gamma_s(s) w_t^2 ds \right), \tag{5.176}$$

where $\gamma_s(s)$ is explicitly given by

$$\begin{aligned} \gamma_s(s) \quad = \quad & \frac{8s-4}{4p_0 h(s)} \left[\arctan\left(\frac{2s-1}{2p_0}\right) + p_0 \ln\left(2s - 1 + h(s)\right) \right. \\ & \left. - \arctan\frac{1}{2p_0} + p_0 \ln\left(\sqrt{4p_0^2+1} - 1\right) \right] \\ & + \frac{h(s)}{2p_0} \left(\frac{1}{p_0 \left(1 + \frac{(2s-1)^2}{4p_0^2}\right)} + \frac{p_0 \left(2 + \frac{8s-4}{2h(s)}\right)}{2s - 1 + h(s)} \right). \end{aligned}$$

It can be shown that $\gamma_s(s) \geq 1$ for $s \in [0, 1]$. Eq. (5.176) simplifies to

$$A_1 \leq \beta \left(\gamma(1) w_t^2(1) - \int_0^1 w_t^2 ds \right). \tag{5.177}$$

Integration of Eq. (5.174) by parts produces

$$A_2 = 2\beta\left(\gamma(1)p(1)w_s^2(1) - \int_0^1 \gamma_s(s)p(s)w_s^2 ds - \int_0^1 \gamma(s)p(s)w_s w_{ss} ds\right), \qquad (5.178)$$

Eq. (5.174) also expands to

$$A_2 = 2\beta\left(\int_0^1 \gamma(s)w_s^2 p_s(s)ds + \int_0^1 \gamma(s)p(s)w_s w_{ss} ds\right). \qquad (5.179)$$

Combination of Eqs. (5.178) and (5.179) simplifies

$$A_2 = \beta\left(\gamma(1)p(1)w_s^2(1) - \int_0^1 \left[\gamma_s(s)p(s) - \gamma(s)p_s(s)\right] w_s^2 ds\right). \qquad (5.180)$$

From the structure of $\gamma(s)$ and $p(s)$ we have

$$\gamma_s(s)p(s) - \gamma(s)p_s(s) = 1 + p(s). \qquad (5.181)$$

Substitution of Eq. (5.181) into Eq. (5.180) produces

$$A_2 \leq \beta\left(\gamma(1)p(1)w_s^2(1) - \int_0^1 p(s)w_s^2 ds\right). \qquad (5.182)$$

Substitution of Eqs. (5.177) and (5.182) into Eq. (5.173) and then substituting the resulting expression along with Eq. (5.172) into Eq. (5.168), we have

$$\begin{aligned} \dot{V} \;\leq\; & -\left(\frac{p(1)}{2} - \beta\gamma(1)\right) w_t^2(1) - \left(\frac{p(1)}{2} - \beta\gamma(1)p(1)\right) w_s^2(1)) \\ & -\left(k_s - \frac{p(1)}{2}\right)\eta^2 - 2\beta E_s, \end{aligned} \qquad (5.183)$$

where Eq. (5.160) has been used. From Eq. (5.183), it is clear that if the controller gain k_s is selected according to Eq. (5.155) and the design constant β is selected according to

$$\beta < \min\left\{\frac{p(1)}{2\gamma(1)}, \frac{1}{2\gamma(1)}, \frac{\min\{1, p\}}{4\gamma(1)}\right\}, \qquad (5.184)$$

then $\dot{V}(t)$ can be upper bounded by the nonpositive scalar function given in Eq. (5.167). □

From (5.163) and (5.167), we obtain the following upper bound for the time derivative of $V(t)$

$$\dot{V}(t) \leq -\frac{\lambda_3}{\lambda_2} V(t), \tag{5.185}$$

whose maximal solution yields

$$V(t) \leq V(0) \exp\left(-\frac{\lambda_3}{\lambda_2} t\right) \leq \lambda_2 \kappa_o \exp\left(-\frac{\lambda_3}{\lambda_2} t\right), \tag{5.186}$$

using Eq. (5.163) with

$$\kappa_o = \frac{1}{2} \int_0^1 w_t^2(s,0) ds + \frac{1}{2} \int_0^1 p(s) w_s^2(s,0) ds + (w_t(1,0) + w_s(1,0))^2. \tag{5.187}$$

In addition, we use Eqs. (3.5), (5.160), and (5.163) to formulate

$$\frac{p_0}{2} w^2(x,t) \leq \frac{1}{2} \int_0^1 p(s) w_s^2 ds \leq E_s(t) \leq \frac{1}{\lambda_1} V(t) \qquad \forall x \in [0,1]. \tag{5.188}$$

The inequality given in Eq. (5.154) now directly follows by combining Eqs. (5.186) and (5.188) and then using Eqs. (5.156) and (5.160). Under the boundary controlled string assumptions, we can also show that all closed loop signals are bounded.
□

5.1.6 Rigid Body Setpoint Regulation

The previous boundary control examples involve systems without rigid body modes. In open loop, the systems are constrained from translation and rotation by geometric boundary conditions. In many applications, however, the open loop system can translate and/or rotate without vibration. For example, rigid body modes occur in a single flexible link robot arm and a spacecraft with flexible solar arrays. The control objective in these cases is to regulate the system output displacement at a desired point *and* asymptotically stabilize the system vibration. In the previous chapter, LaSalle's Invariance Principle is used to prove that passive feedback using one actuator can stabilize many of these free boundary problems. Passive control, however, may not provide the desired performance. To meet these two objectives with an EMK controller, we require two actuators for a 1D continua, one on each boundary. In this section, we derive and experimentally test an EMK rigid body translation and vibration controller for an axially moving material system.

Tension and Speed Setpoint Regulation for Axially Moving Materials

Recall the axially moving material model for longitudinal vibration, translation, and tension governed by Eqs. (2.41) - (2.43) and shown in Fig. 6 of Chapter 2 . We now design torque controllers for the two rollers that enable tension and speed setpoint regulation for the material.

To facilitate the control development, we introduce the variables

$$y(t) = u(L,t) \qquad \text{and} \qquad v(x,t) = y(t) - u(x,t) + \frac{P_D}{EA}(x - L), \tag{5.189}$$

where P_D is the constant desired material tension. We note that $EAv_x(x,t) = P(t) - P_D$ is the tension error where the material tension $P(x,t) = EAu_x(x,t)$. If we differentiate Eq. (5.189) with respect to time we obtain $v_t(x,t) = \dot{y} - u_t(x,t)$. Thus, if we design a controller to force $\dot{y} \to v_d$, $v_x(x,t) \to 0$, and $v_t(x,t) \to 0$ for all $x \in [0, L]$ as $t \to \infty$, then we ensure the entire web asymptotically moves at the desired speed v_d with the desired tension P_D.

Eq. (5.189) transforms the field Eq. (2.41) to

$$\rho v_{tt}(x,t) - EAv_{xx}(x,t) = \rho \ddot{y}(t) \tag{5.190}$$

and the boundary conditions Eqs. (2.42) – (2.43) to

$$v(L,t) = 0 \tag{5.191}$$

$$m v_{tt}(0,t) - EAv_x(0,t) + P_D = m\ddot{y}(t) - f_0(t). \tag{5.192}$$

The right roller dynamics ($x = L$) become

$$m\ddot{y}(t) - EAv_x(L,t) + P_D = f_L(t). \tag{5.193}$$

Controller Derivation We present the structure of a model-based boundary control law that exponentially regulates the tension and speed to desired setpoints for the system mathematically described by Eqs. (5.190) through (5.193). First, we define the axial speed setpoint error

$$\eta_1(t) = v_d - \dot{y}(t) = v_d - u_t(L,t) \tag{5.194}$$

and tension setpoint error

$$\eta_2(t) = v_t(0,t) - \kappa v_x(0,t) = u_t(L,t) - u_t(0,t) - \frac{\kappa}{EA}(P_D - P(0,t)), \tag{5.195}$$

where κ is a positive scalar control gain. The roller torque controllers drive $\eta_1(t)$ and $\eta_2(t)$ to zero as $t \to \infty$ to enforce speed and tension setpoint regulation and damp axial vibration. To obtain the open-loop dynamics for the axial speed setpoint error $\eta_1(t)$, we first differentiate Eq. (5.194) with respect to time, multiply the resulting expression by m, and substitute Eq. (5.193), to obtain

$$m\dot{\eta}_1(t) = -EAv_x(L,t) + P_D - f_L(t) = -m\ddot{y}(t). \tag{5.196}$$

We define the EMK speed setpoint controller

$$f_L(t) = P(L,t) + k_p\eta_1(t) + k_i\int_0^t \eta_1(\tau)\,d\tau, \tag{5.197}$$

where k_p and k_i are constant, positive, scalar proportional and integral speed gains, respectively. Implementation of the speed control law (see Eq. (5.197)) requires the measurement of the roller speed and tension at the right boundary. This can be achieved with a tachometer or an encoder on the right roller and a tension sensor near the right roller. Substitution of Eq. (5.197) into Eq. (5.196), yields the closed-loop dynamics

$$m\dot{\eta}_1(t) = -m\ddot{y}(t) = -k_p\eta_1(t) - k_i\int_0^t \eta_1(\tau)\,d\tau. \tag{5.198}$$

Differentiation of Eq. (5.195) with respect to time, multiplication by m, and substitution of Eq. (5.192), produces

$$m\dot{\eta}_2(t) = [m\ddot{y}(t) + EAv_x(0,t) - P_D - f_0(t)] - m\kappa v_{xt}(0,t). \tag{5.199}$$

We define the tension setpoint controller

$$f_0(t) = \frac{m\kappa}{EA}P_t(0,t) - P(0,t) + k_{s2}\eta_2 + k_p\eta_1 + k_i\int_0^t \eta_1(\tau)\,d\tau, \tag{5.200}$$

where Eq. (5.192) has been used and k_{s2} is a positive constant control gain. Implementation of the tension control law Eq. (5.200) can be achieved with a tachometer or encoder on the left roller and a tension sensor near the left roller. Filtered backwards difference of the tension signal can provide $P_t(0,t)$. Note that exact knowledge of m, EA, P_{B0}, and P_{BL} (see Eq. (2.45)) are required for control implementation. Substitution of Eq. (5.200) into Eq. (5.199) yields the closed-loop dynamics

$$m\dot{\eta}_2(t) = -k_{s2}\eta_2(t). \tag{5.201}$$

Figure 5 shows a block diagram of the proposed control scheme. The proportional plus integral (PI) speed controller includes a tension feedforward term that cancels the material tension applied to the roller. The proportional plus derivative (PD) tension controller includes tension and speed feedforward terms as well as a relative speed damping term.

Theorem 50 *The speed and tension setpoint controllers given in Eqs. (5.197) and (5.200), respectively, exponentially regulate the speed, tension, and vibration as follows*

$$\begin{aligned} |\eta_1(t)| &\le \alpha_1 \exp(-\gamma_1 t) \\ |v(x,t)| &\le \alpha_2 \exp(-\gamma_2 t), \end{aligned} \tag{5.202}$$

where $\alpha_1, \gamma_1, \alpha_2$ and γ_2 are positive constants.

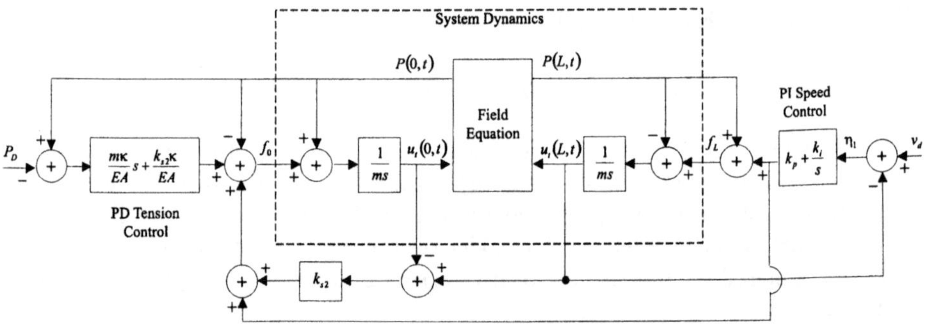

Figure 5 Tension/speed control block diagram

Proof: We first define the Lyapunov functional

$$V(t) = E_s(t) + E_c(t) + \frac{1}{2}m\eta_2^2(t) + \frac{1}{2}z^T(t)\,Tz(t), \tag{5.203}$$

where

$$E_s(t) = \frac{1}{2}\beta_s\rho\int_0^L v_t^2 ds + \frac{1}{2}\beta_s EA\int_0^L v_s^2 ds, \tag{5.204}$$

$$E_c(t) = 2\beta_c\rho\int_0^L (s-L)\,v_t v_s ds, \tag{5.205}$$

$$T = \begin{bmatrix} \frac{k_i}{m} + \frac{k_p}{2m} & \frac{1}{2} \\ \frac{1}{2} & 1 \end{bmatrix}, \tag{5.206}$$

$$z(t) = \begin{bmatrix} \int_0^t \eta_1(\tau)\,d\tau & \eta_1(t) \end{bmatrix}^T, \tag{5.207}$$

with small, positive constants β_s and β_c. After selecting β_c sufficiently small and $k_p > \frac{m}{2}$, the function $V(t)$ defined in Eq. (5.203) can be upper and lower bounded as follows

$$\lambda_1\left(E_s(t) + \eta_2^2(t) + \|z(t)\|^2\right) \leq V(t) \leq \lambda_2\left(E_s(t) + \eta_2^2(t) + \|z(t)\|^2\right), \tag{5.208}$$

where λ_1 and λ_2 are positive constants.

Lemma 51 *The function given in Eq. (5.203) can be upper and lower bounded as shown in Ineq. (5.208).*

Proof: The function $E_s(t)$, defined in Eq. (5.204), can be lower bounded as follows

$$E_s(t) \geq \frac{\beta_s}{2}\min\{\rho, EA\}\int_0^L \left(v_t^2 + v_s^2\right) ds. \tag{5.209}$$

In addition, the function $E_c(t)$, defined in Eq. (5.205), can be upper bounded as follows

$$E_c \leq 2\beta_c \rho L \int_0^L \left(v_t^2 + v_s^2\right) ds, \tag{5.210}$$

where Ineq. (3.10) has been used. From Ineqs. (5.209) and (5.210), we obtain the following upper and lower bounds for $E_c(t)$

$$-\frac{4\beta_c\rho L}{\beta_s \min\{\rho, EA\}}E_s(t) \leq E_c(t) \leq \frac{4\beta_c\rho L}{\beta_s \min\{\rho, EA\}}E_s(t). \tag{5.211}$$

From the structure of Ineq. (5.211), we can state the following inequality

$$0 \leq \xi_1 E_s \leq E_s + E_c \leq \xi_2 E_s, \tag{5.212}$$

where the positive scalar constants

$$\xi_1 = \left(1 - \frac{4\beta_c\rho L}{\beta_s \min\{\rho, EA\}}\right) > 0 \quad \text{and} \quad \xi_2 = \left(1 + \frac{4\beta_c\rho L}{\beta_s \min\{\rho, EA\}}\right) > 1 \tag{5.213}$$

provided β_c is selected to satisfy the inequality

$$\beta_c < \frac{\beta_s \min\{\rho, EA\}}{4\rho L}. \tag{5.214}$$

To ensure that the matrix T defined in Eq. (5.206) remains positive definite, we require

$$k_p > \frac{m}{2}. \tag{5.215}$$

From the structure of Eq. (5.203) and Ineq. (5.212) and after satisfying Ineqs. (5.214) and (5.215), we can obtain the upper and lower bounds for $V(t)$ in Ineq. (5.208) with

$$\begin{aligned}
\lambda_1 &= \min\left\{\frac{1}{2}\lambda_{\min}\{P\}, \frac{m}{2}, 1 - \frac{4\beta_c\rho L}{\beta_s \min\{\rho, EA\}}\right\} > 0 \\
\lambda_2 &= \max\left\{\frac{1}{2}\lambda_{\max}\{P\}, \frac{m}{2}, 1 + \frac{4\beta_c\rho L}{\beta_s \min\{\rho, EA\}}\right\} > 1.
\end{aligned} \tag{5.216}$$

□

After differentiation of Eq. (5.203) with respect to time and use of the following Lemma, we obtain an upper bound for the time derivative of $V(t)$ as follows

$$\dot{V}(t) \leq -\lambda_3\left(E_s(t) + \eta_2^2(t) + \|z(t)\|^2\right) + \bar{\alpha}\exp(-\bar{\gamma}t), \tag{5.217}$$

where $\lambda_3, \bar{\alpha}$ and $\bar{\gamma}$ are positive constants defined in the Lemma.

Lemma 52 *The time derivative of the function $V(t)$ defined in Eq. (5.203) can be upper bounded as shown in Ineq. (5.217).*

Proof: After differentiating Eq. (5.203) with respect to time and substituting Eqs. (5.198) and (5.201) and the time derivative of Eq. (5.207), we have

$$\dot{V}(t) = \dot{E}_s(t) + \dot{E}_c(t) - z^T(t) Q z(t) - k_{s2}\eta_2^2(t). \tag{5.218}$$

The matrix Q in Eq. (5.218) is defined as follows

$$Q = \begin{bmatrix} \frac{k_i}{2m} & 0 \\ 0 & \frac{k_p}{m} - \frac{1}{2} \end{bmatrix} = Q^T > 0 \tag{5.219}$$

if Ineq. (5.215) is satisfied. To determine $\dot{E}_s(t)$ in Eq. (5.218), we differentiate Eq. (5.204) with respect to time, substitute the field Eq. (5.190), and cancel common terms to obtain

$$\dot{E}_s = \beta_s \rho \int_0^L \ddot{y} v_t ds + \left[\beta_s EA \int_0^L v_{ss} v_t ds\right] + \beta_s EA \int_0^L v_s v_{st} ds. \tag{5.220}$$

After integrating the bracketed term in Eq. (5.220) by parts once, applying the boundary conditions given in Eq. (5.191), and cancelling common terms, we obtain

$$\dot{E}_s = \beta_s \rho \int_0^L \ddot{y} v_t ds - \beta_s EA v_t(0) v_x(0), \tag{5.221}$$

which can be rewritten as follows

$$\dot{E}_s = \left[\beta_s \rho \int_0^L \ddot{y} v_t ds\right] - \beta_s \frac{EA}{2\kappa} v_t^2(0) - \beta_s \frac{EA\kappa}{2} v_x^2(0) + \beta_s \frac{EA}{2\kappa} \eta_2^2, \tag{5.222}$$

where Eq. (5.195) has been used.

To determine $\dot{E}_c(t)$, we first differentiate Eq. (5.205) with respect to time and then substitute the field Eq. (5.190) to obtain

$$\begin{aligned} \dot{E}_c &= 2\beta_c \rho \int_0^L (s-L)\ddot{y} v_s ds + \left[2\beta_c EA \int_0^L (x-L) v_{ss} v_s ds\right] \\ &\quad + \left[2\beta_c \rho \int_0^L (s-L) v_t v_{st} ds\right]. \end{aligned} \tag{5.223}$$

After integrating the bracketed terms in Eq. (5.223) by parts, and then applying the boundary conditions of Eq. (5.191), we obtain

$$\begin{aligned}\dot{E}_c &= 2\beta_c EALv_x^2(0) + 2\beta_c\rho Lv_t^2(0) + 2\beta_c\rho\int_0^L (s-L)\ddot{y}v_s ds \\ &-2\beta_c EA\int_0^L v_s^2 ds - 2\beta_c EA\int_0^L (s-L)v_s v_{ss} ds \\ &-2\beta_c\rho\int_0^L v_t^2 ds - 2\beta_c\rho\int_0^L (s-L)v_t v_{st} ds. \end{aligned} \quad (5.224)$$

After adding Eqs. (5.223) and (5.224) and then dividing the resultant equation by 2, we obtain

$$\begin{aligned}\dot{E}_c &= \beta_c EALv_x^2(0) + \beta_c\rho Lv_t^2(0) + 2\beta_c\rho\int_0^L (s-L)\ddot{y}v_s ds \\ &-\beta_c EA\int_0^L v_s^2 ds - \beta_c\rho\int_0^L v_t^2 ds. \end{aligned} \quad (5.225)$$

Substituting Eqs. (5.222) and (5.225) into Eq. (5.218) produces

$$\begin{aligned}\dot{V} &= -\left(\frac{\beta_s EA}{2\kappa} - \beta_c\rho L\right)v_t^2(0) - \left(\frac{\beta_s EA\kappa}{2} - \beta_c EAL\right)v_x^2(0) - \left(k_{s2} - \frac{\beta_s EA}{2\kappa}\right)\eta_2^2 \\ &-z^T Qz - \beta_c EA\int_0^L v_s^2 ds - \beta_c\rho\int_0^L v_t^2 ds \\ &+\left[\beta_s\rho\int_0^L \ddot{y}v_t ds\right] + \left[2\beta_c\rho + \int_0^L (s-L)\ddot{y}v_s ds\right]. \end{aligned} \quad (5.226)$$

If the weighting constants β_c and β_s are chosen according to

$$\beta_c < \min\left\{\frac{\beta_s EA}{2\rho L\kappa}, \frac{\beta_s\kappa}{2L}\right\}, \qquad \beta_s < \frac{2k_{s2}\kappa}{EA}, \quad (5.227)$$

then application of Ineq. (3.12) to the bracketed terms in Eq. (5.226) produces

$$\begin{aligned}\dot{V} &\leq -\left(k_{s2} - \frac{\beta_s EA}{2\kappa}\right)\eta_2^2 - z^T Qz - \left(\beta_c EA - 2\beta_c\rho L\delta_2\right)\int_0^L v_s^2 ds \\ &- \left(\beta_c\rho - \rho\beta_s\delta_1\right)\int_0^L v_t^2 ds + \left[\int_0^L \ddot{y}^2\left(\frac{\beta_s\rho}{\delta_1} + \frac{2\beta_c\rho L}{\delta_2}\right)ds\right]. \end{aligned} \quad (5.228)$$

From Eq. (5.198), it is obvious that $\dot{y}(t), \ddot{y}(t)$ are exponentially driven to zero so the bracketed term in Ineq. (5.228) can be simplified as follows

$$\int_0^L \ddot{y}^2\left(\frac{\beta_s\rho}{\delta_1} + \frac{2\beta_c\rho L}{\delta_2}\right)ds = \ddot{y}^2\left(\frac{\beta_s\rho}{\delta_1} + \frac{2\beta_c\rho L}{\delta_2}\right)L \leq \bar{\alpha}\exp(-\bar{\gamma}t), \quad (5.229)$$

where $\bar{\alpha}, \bar{\gamma}$ are positive constants. Substitution of Eq. (5.229) into Eq. (5.228) produces

$$\dot{V} \leq -2\frac{\beta_c}{\beta_s}\left[\left(1-\frac{2\rho L\delta_2}{EA}\right)\frac{\beta_s EA}{2}\int_0^L v_s^2 ds + \left(1-\frac{\beta_s\delta_1}{\beta_c}\right)\frac{\beta_s\rho}{2}\int_0^L v_t^2 ds\right] - \left(k_{s2}-\frac{\beta_s EA}{2\kappa}\right)\eta_2^2 - z^T Q z + \bar{\alpha}\exp(-\bar{\gamma}t). \tag{5.230}$$

If the constants δ_1 and δ_2 in Eq. (5.230) are chosen to satisfy

$$\delta_1 < \frac{\beta_c}{2\beta_s}, \quad \delta_2 < \frac{EA}{4\rho L}, \tag{5.231}$$

we can rewrite Eq. (5.230) as follows After applying the definition of $E_s(t)$ to the bracketed term in Eq. (5.232

$$\dot{V} \leq -\frac{\beta_c}{\beta_s}\left[\frac{1}{2}\beta_s EA\int_0^L v_s^2 ds + \frac{1}{2}\beta_s\rho\int_0^L v_t^2 ds\right] - \left(k_{s2}-\frac{\beta_s EA}{2\kappa}\right)\eta_2^2 - z^T Q z + \bar{\alpha}\exp(-\bar{\gamma}t). \tag{5.232}$$

The definition of

$$\lambda_3 = \min\left\{\frac{\beta_c}{\beta_s}, k_{s2}-\frac{\beta_s EA}{2\kappa}, \lambda_{\min}\{Q\}\right\} \tag{5.233}$$

and satisfaction of Ineqs. (5.227) and

$$0 < \beta_c < \min\left\{\frac{\beta_s EA}{2\rho L\kappa}, \frac{\beta_s \kappa}{2L}, \frac{\beta_s \min\{\rho, EA\}}{4\rho L}\right\} \tag{5.234}$$

produce the upper bound shown in Ineq. (5.217).
□

From Ineqs. (5.208) and (5.217), we can formulate the following upper bound on the solution of $V(t)$

$$V(t) \leq V(0)\exp\left(-\frac{\lambda_3}{\lambda_2}t\right) + \bar{a}\exp\left(-\bar{b}t\right), \tag{5.235}$$

where $\bar{a}$ and $\bar{b}$ are positive constants. In addition, we can use Ineqs. (5.208) and (5.235) to formulate the following inequalities

$$\frac{\beta_s EA}{2L}v^2(x,t) \leq \frac{\beta_s EA}{2}\int_0^L v_s^2 ds \leq E_s(t) \leq \frac{1}{\lambda_1}V(t)\ \forall x \in [0, L], \tag{5.236}$$

where Ineq. (3.5) is used. Substituting for $V(t)$ in Eq. (5.236) from Eq. (5.235), we obtain the following upper bound for $v^2(x,t)$

$$v^2(x,t) \leq \frac{2L}{\lambda_1 EA\beta_s}\left[V(0)\exp\left(-\frac{\lambda_3}{\lambda_2}t\right) + \bar{a}\exp\left(-\bar{b}t\right)\right]. \tag{5.237}$$

Thus, $v^2(x,t)$ decays exponentially to zero yielding the second result in Eq. (5.202). The first result in Eq. (5.202) follows directly from the structure of Eq. (5.198).
□

The two control objectives are to drive the material tension $P(x,t)$ to a constant desired tension setpoint P_D and the material speed to a constant desired speed setpoint v_d. From the first inequality in Ineq. (5.202), we know that $\dot{y}(t)$ is exponentially driven to v_d. Unfortunately, we cannot differentiate Ineq. (5.237) with respect to space and time to show that to show that $v_t(x,t)$ and $v_x(x,t)$ exponentially decay to zero, respectively. From Eq. (5.203) and Ineq. (5.235), however, we show that E_s exponentially decays to zero. It seems reasonable to assume that if the norms of the velocity and strain distributions (kinetic and potential energies) decay exponentially to zero then the pointwise velocity and strain should also decay exponentially to zero. If we assume this to be true, the time derivative of Eq. (5.189) yields the velocity setpoint regulation objective (i.e., $u_t(x,t)$ is exponentially driven to v_d) and the spatial derivative of Eq. (5.189) yields the tension setpoint regulation objective (*i.e.*, $EAu_x(x,t)$ is exponentially driven to P_D).

Experimental Setup To test the stability and performance of the proposed controller we develop a small-scale experiment. A schematic diagram of the experimental setup used to implement the controller is shown in Figure 6. The experimental test-stand consists of an elastic rubber belt moving axially over two pulleys actuated by brushed DC motors. The four tension sensors and roller assemblies laterally position the moving material and provide measurements of the forward boundary tensions $P(0,t)$ and $P(L,t)$ and the back boundary tensions P_{B0} and P_{BL} used by the controller. The brushed DC motors have 1024 counts per revolution encoders that measure the angular displacements of the rotors. The Transducer Techniques' MLP-10 tension sensors (measurement range: ± 40 [N]) are positioned to ensure maximum surface contact between the material and the pulleys in order to prevent slip. The angles subtended by the material as it approaches and leaves the tension sensor are set equal to produce a resultant force along the measurement axis of the tension sensor, avoiding potential damage to the tension sensor from an induced bending moment.

A Pentium 266 MHz PC running QNX hosts the control algorithm. *Qmotor* [104] provides an environment to write the 'C' control program. The *MultiQ* I/O board provides for the data transfer between computer subsystem and the electrical interface. Two A/D channels are used to sense the currents flowing through the stator

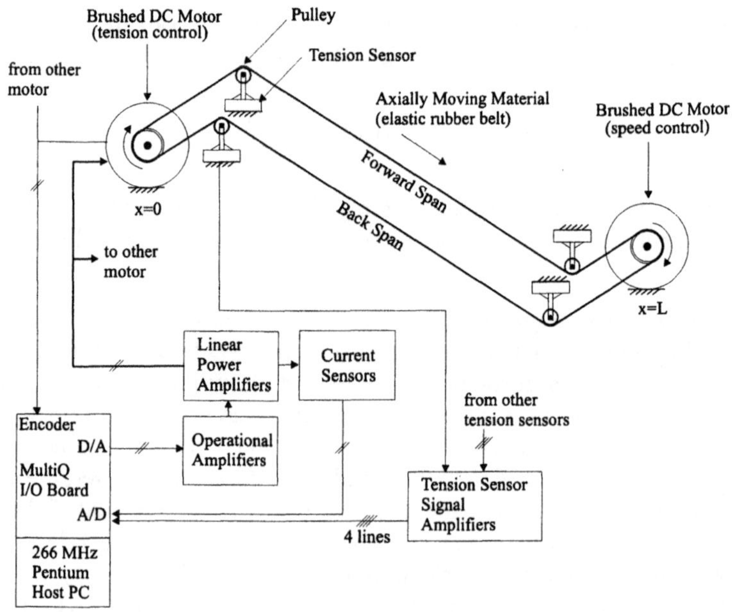

Figure 6 Schematic diagram of the tension and speed control experiment

windings of the DC motors, and four additional A/D channels are used to read in the tension sensor signals. Two D/A channels output voltages that drive the DC motors. These voltages go through two stages of amplification, consisting of OP07C operational amplifiers and Techron linear power amplifiers, for the first and second stages, respectively. High-gain current feedback ensures that the required torque is applied to the motors. All controllers are implemented using a sampling period of 0.5 [m-sec]. The axial speed of the material and the derivative of the boundary tension at $x = 0$ are obtained using a filtered backwards difference algorithm. The following parameters associated with the system have been obtained using standard test procedures and engineering handbook tables: $r_0 = r_L = 0.0287$ [m], $EA = 12.06$ [N], $m = 0.0346$ [kg], $J_0 = J_L = 2.8353 \times 10^{-5}$ $[\text{kg} - \text{m}^2]$, and $L = 1.42$ [m].

Experimental Results The objective of the experiment is to regulate the material tension at 8.0 [N] and move the material along a smooth, exponentially stepped, desired axial speed setpoint trajectory. In order to mimic real-world industrial processes (such as high-speed label printing), the desired speed of the material is aggressively driven to 0.75 [m/s] and back to 0 [m/s] within a time duration of 0.5 [sec] and is repeated every 10 [sec]. Process-line disturbances leading to a sudden change in material tension are also simulated by applying a constant reverse torque on the motor at $x = 0$ for a duration of 0.5 [sec] at 10 [sec] intervals. The control is initially applied after a one second time delay.

First, a PI speed controller consisting of a constant open-loop bias torque to the motor at $x = 0$, and PI feedback of the material speed at $x = L$, is implemented as follows

$$\begin{aligned} f_0(t) &= 15.46 \text{ [N]} \\ f_L(t) &= k_p \eta_1(t) + k_i \int_0^t \eta_1(\tau)\, d\tau \text{ [N]}, \end{aligned} \tag{5.238}$$

with best-performing control gains

$$k_p = 30 \text{ and } k_i = 100.$$

The speed controller $f_L(t)$ has the same structure as the controller given in Eq. (5.197) without the feedforward boundary tension term, $P(L,t)$. Figure 7 shows the forward span boundary tensions, the material speed measured at the two boundaries and the torque applied to the two motors. The PI speed controller does a good job of regulating $u_t(L,t)$ despite the start-stop trajectories at $t = 10$ and 20 [sec] and the torque disturbances at $t = 5, 15$ and 25 [sec]. The speed $u_t(0,t)$ and tension, however, do not track their desired values.

Second, proportional (P) tension feedback is added to the PI speed controller in Eq. (5.238)

$$f_0(t) = 10.03 + k_t (P_D - P(0,t)) \text{ [N]}, \tag{5.239}$$

where k_t is a positive constant, proportional gain. The best performing gains are

$$k_t = 500,\ k_p = 34, \text{ and } k_i = 650.$$

Figure 8 shows the forward span boundary tensions, the material speed measured at the two boundaries and the torque applied to the two motors. The addition of P tension feedback greatly improves the tension setpoint regulation. The tension and speed, however, remain sensitive to the torque and speed disturbances.

Finally, the full-order controller given by Eqs. (5.197) and (5.200) is implemented. The performance of the full-order controller shown in Figure 9 uses the following control gains:

$$k_p = 40,\ k_i = 600,\ k_{s2} = 26, \text{ and } \kappa = 40.$$

The error between the desired and actual material speeds at $x = 0$ and $x = L$ for the full-order controller response is very small (peak error is only 0.07 [m/s]). Thus, the desired and actual speed signals completely overlap each other in Figure 9. The forward span boundary tensions, the material boundary velocities and the torques applied to the two motors are also plotted. The full-order controller does an excellent job of regulating the speed and tension throughout the span.

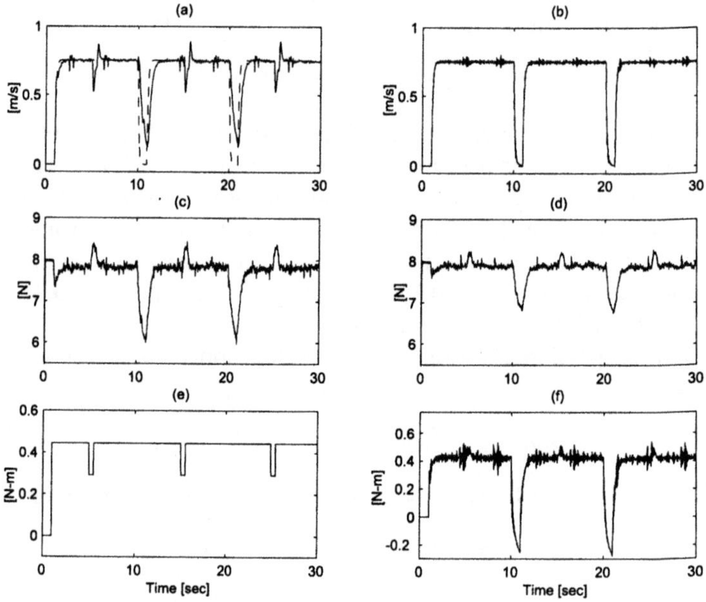

Figure 7 PI-speed controller without tension feedback at $x = 0$: (a) Material speed at $x = 0$, $u_t(0,t)$ (solid line), desired material speed (dotted line), (b) Material speed at $x = L$, $u_t(L,t)$ (solid line), desired material speed (dotted line), (c) Material tension at $x = 0$, $P(0,t)$, (d) Material tension at $x = L$, $P(L,t)$, (e) Torque applied to the motor at $x = 0$, $\tau_0(t)$, and (f) Torque applied to the motor at $x = L$, $\tau_L(t)$.

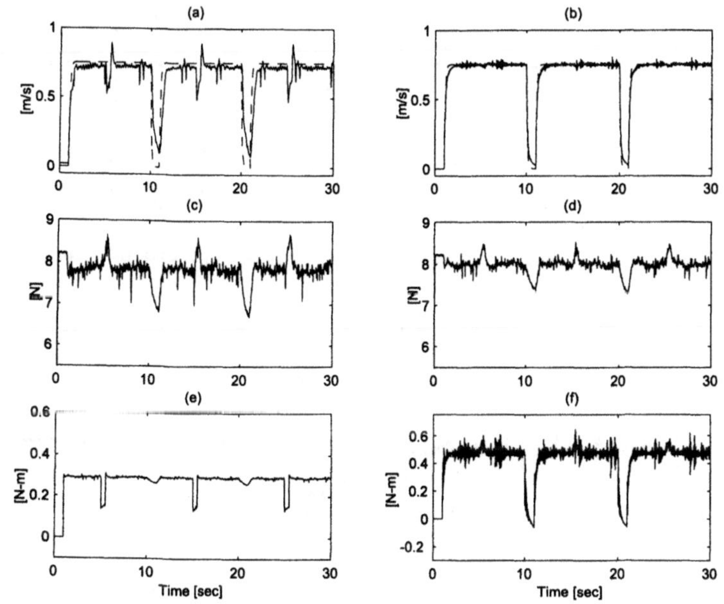

Figure 8 PI-speed controller with tension feedback at $x = 0$: (a) Material speed at $x = 0$, $u_t(0,t)$ (solid line), desired material speed (dotted line), (b) Material speed at $x = L$, $u_t(L,t)$ (solid line), desired material speed (dotted line), (c) Material tension at $x = 0$, $P(0,t)$, (d) Material tension at $x = L$, $P(L,t)$, (e) Torque applied to the motor at $x = 0$, $\tau_0(t)$, and (f) Torque applied to the motor at $x = L$, $\tau_L(t)$.

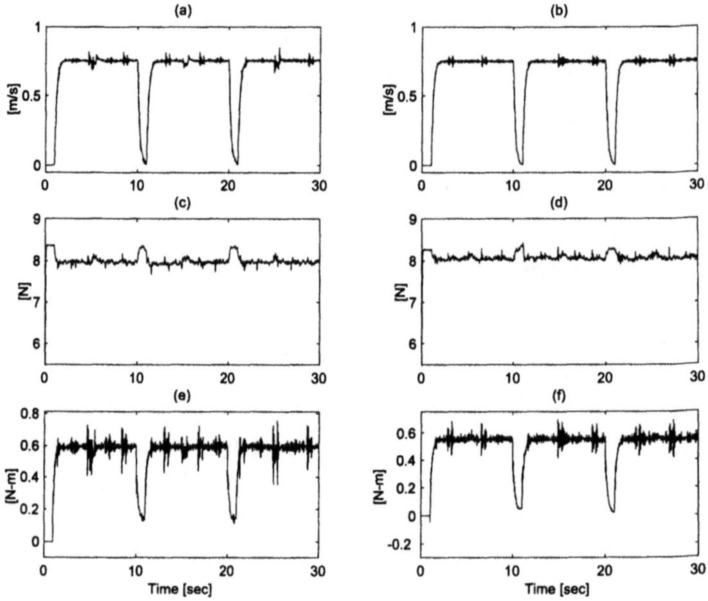

Figure 9 Full-order controller: (a) Material speed at $x = 0$, $u_t(0,t)$ (solid line), desired material speed (dotted line), (b) Material speed at $x = L$, $u_t(L,t)$ (solid line), desired material speed (dotted line), (c) Material tension at $x = 0$, $P(0,t)$, (d) Material tension at $x = L$, $P(L,t)$, (e) Torque applied to the motor at $x = 0$, $\tau_0(t)$, and (f) Torque applied to the motor at $x = L$, $\tau_L(t)$.

The experimental results show that the maximum speed setpoint error at $x = 0$ with both PI speed controllers is approximately 3 times that of the full-order controller in response to the torque disturbance and nearly an order of magnitude higher in response to the aggressive start-stop speed trajectory. However, at $x = L$, the maximum speed setpoint errors of both PI speed controllers and full-order controller are comparable for both disturbances. The full-order controller, however, has half the speed settling time of the other controllers. The maximum tension setpoint error at $x = 0$ with both PI speed controllers is nearly twice that of the full-order controller in response to the torque disturbance. When an aggressive start-stop desired speed trajectory is commanded, the PI speed controller with tension feedback fares much better than its counterpart without tension feedback. Specifically, the maximum tension regulation error is lowered by nearly 70% with the introduction of tension feedback at $x = 0$. The full-order controller, however, out-performs the PI speed controller with tension feedback by a factor of 3. At $x = L$, the tension setpoint regulation in response to the back torque disturbance is nearly 50% lower with the full-order controller than both PI speed controllers. With a start-stop speed disturbance, the full-order controller improves tension setpoint regulation by a factor of 3 over the PI speed controller without tension feedback and a factor of 2 over the PI speed controller with tension feedback.

5.2 Domain Control

Domain control involves actuators inside the boundaries of a distributed system. In the previous chapter, we show that a carefully located sensor and actuator with passive feedback can asymptotically stabilize many 1D continua. Unfortunately, this configuration does not lend itself to Lyapunov-based derivations of EMK and Adaptive controllers that provide good performance and stability robustness. A domain actuator effectively divides the domain into two subdomains. To asymptotically stabilize the vibration in both subdomains requires two actuators. Alternatively, one actuator can be used to asymptotically stabilize one subdomain despite bounded disturbances in the other subdomain. In this section, we derive acoustic noise and axially moving beam vibration isolators based on this principle.

5.2.1 Acoustic Noise Isolator

We now refer to the system model shown in Figure 5 of Chapter 2 and described by Eqs. (2.35) - (2.39). The duct is divided into a quiet or controlled subdomain for $x \in (0, a)$ and a noisy or uncontrolled subdomain for $x \in (a, L)$ by the speaker actuator located at $x = a$. The control objectives are to (i) asymptotically regulate the quiet subdomain acoustic displacement to zero; (ii) maintain boundedness of the noisy subdomain acoustic displacement; and (iii) insure boundedness of the control force despite unknown but bounded disturbances in the noisy subdomain. Heuristically, the control objective is to turn the control speaker into a one-way acoustic valve that blocks the "flow" of noise from the noisy subdomain to the quiet subdomain but facilitates noise flow in the opposite direction.

Boundedness of the Noisy Subdomain

First, we show that if the control algorithm ensures that the speaker displacement is bounded (*i.e.* $u(a,t) \in \mathcal{L}_\infty$), then the noisy subdomain response is bounded.

Lemma 53 *If $f_u(t) \in \mathcal{L}_\infty$ and $p_l(t) \in \mathcal{L}_\infty$ and $u(a,t) \in \mathcal{L}_\infty$ up to second time derivatives, then $u(x,t) \in \mathcal{L}_\infty([a,L])$.*

Proof: Substitution of

$$v(x,t) = u(x,t) - u(a,t) - \frac{x-a}{\mathcal{B}} p(L,t) \tag{5.240}$$

into Eq. (2.35) yields

$$\rho_0 v_{tt} - \tau v_{xxt} - \mathcal{B} v_{xx} = F_u(x,t) \text{ for } x \in (a,L), \tag{5.241}$$

with the boundary conditions

$$v(a,t) = v_x(L,t) = 0, \tag{5.242}$$

where

$$F_u(x,t) = f_u(x,t) - \rho_0 u_{tt}(a,t) - \rho_0 \frac{x-a}{\mathcal{B}} p_{tt}(L,t). \tag{5.243}$$

Note that $F_u \in \mathcal{L}_\infty([a,L])$ by the Lemma assumptions.

A Lyapunov functional for the uncontrolled subdomain is chosen to be

$$V_u = E_b^+(t) + E_c^+(t), \tag{5.244}$$

where

$$E_b^+ = \frac{1}{2}\rho_0 \int_a^L v_t^2 dx + \frac{1}{2}\mathcal{B} \int_a^L v_x^2 dx, \tag{5.245}$$

and

$$E_c^+ = \beta \rho_0 \int_a^L v v_t dx \tag{5.246}$$

with β a small positive constant. Use of Ineqs. (3.16) and (3.10) in Eq. (5.246) yields

$$\begin{aligned} \left|E_c^+\right| &\leq \beta\rho_0 \int_a^L \left(v^2 + v_t^2\right) dx \leq \beta\rho_0 \int_a^L \left(v_t^2 + (L-a)^2 v_x^2\right) dx \\ &\leq 2\beta\rho_0 \frac{\max\left(1,(L-a)^2\right)}{\min\left(\rho_0, \mathcal{B}\right)} E_b^+(t), \end{aligned} \tag{5.247}$$

so

$$0 \leq \lambda_1 E_b^+(t) \leq V_u(t) \leq \lambda_2 E_b^+(t), \tag{5.248}$$

where

$$\lambda_1 = 1 - 2\beta\rho_0 \frac{\max\left(1, (L-a)^2\right)}{\min\left(\rho_0, \mathcal{B}\right)}, \tag{5.249}$$

and

$$\lambda_2 = 1 + 2\beta\rho_0 \frac{\max\left(1, (L-a)^2\right)}{\min\left(\rho_0, \mathcal{B}\right)}. \tag{5.250}$$

If we select $\beta < \frac{\min(\rho_0, \mathcal{B})}{2\rho_0 \max(1, l^2)}$, then $\lambda_1 > 0$ and $\lambda_2 > 1$.

The time derivative of $V_u(t)$

$$\dot{V}_u(t) = \dot{E}_b^+ + \dot{E}_c^+ \tag{5.251}$$

depends on

$$\dot{E}_b^+ = -\tau \int_a^L v_{xt}^2 dx + \int_a^L v_t F_u dx, \tag{5.252}$$

where the field equation (5.241), boundary conditions (5.242), and integration by parts have been used. Application of Ineqs. (3.16) and (3.12) for any $\delta_1 > 0$ to Eq. (5.252) yields

$$\dot{E}_b^+ \leq -\frac{\tau}{2} \int_a^L v_{xt}^2 dx - \left(\frac{\tau}{(L-a)^2} - \delta_1\right) \int_a^L v_t^2 dx + \frac{1}{\delta_1} \int_a^L F_u^2 dx. \tag{5.253}$$

The remaining term in $\dot{V}_u(t)$ is

$$\begin{aligned} \frac{1}{\beta}\dot{E}_c^+ &= \rho_0 \int_a^L \left(v_t^2 + v v_{tt}\right) dx \\ &= \beta \int_a^L \rho_0 v_t^2 dx + \beta \int_a^L v\left(\mathcal{B}v_{xx} + \tau v_{xxt} + F_u\right) dx. \end{aligned} \tag{5.254}$$

The terms in Eq. (5.254) simplify via integration by parts, the boundary conditions, and inequalities as follows

$$\int_a^L v\mathcal{B}v_{xx} dx = -\mathcal{B} \int_a^L v_x^2 dx, \tag{5.255}$$

$$-\tau \int_a^L v v_{xxt} dx \leq \delta_2 \tau \int_a^L v_x^2 dx + \frac{\tau}{\delta_2} \int_a^L v_{xt}^2 dx, \tag{5.256}$$

$$\int_a^L v F_u dx \leq \frac{1}{\delta_3} \int_a^L F_u^2 dx + (L-a)^2 \delta_3 \int_a^L v_x^2 dx, \tag{5.257}$$

where δ_2 and δ_3 are arbitrary positive constants.

Substitution of Eqs. (5.253) – (5.257) into Eq. (5.251) yields

$$\begin{aligned}\dot{V}_u(t) &\leq -\left(\frac{\tau}{(L-a)^2}-\delta_1-\beta\rho_0\right)\int_a^L v_t^2dx+\left(\frac{1}{\delta_1}+\frac{\beta}{\delta_3}\right)\int_a^L F_u^2dx \\ &\quad -\beta\left(\mathcal{B}-\delta_2\tau-(L-a)^2\delta_3\right)\int_a^L v_x^2dx-\left(\frac{\tau}{2}-\frac{\tau}{\delta_2}\beta\right)\int_a^L v_{xt}^2dx \\ &\leq -\lambda_3E_b^+(t)+\varepsilon, \end{aligned} \tag{5.258}$$

where

$$\lambda_3=\min\left(\frac{\frac{\tau}{(L-a)^2}-\delta_1-\beta\rho_0}{\rho_0},\frac{\beta\left(\mathcal{B}-\delta_2\tau-(L-a)^2\delta_3\right)}{\mathcal{B}}\right)>0 \tag{5.259}$$

if β, δ_1, δ_2 and δ_3 are sufficiently small and

$$\varepsilon=\left(\frac{1}{\delta}+\frac{\beta}{\delta_2}\right)\max\left(\int_a^L F_u^2dx\right). \tag{5.260}$$

From $F_u(t)\in\mathcal{L}_\infty([a,L])$ we have $\varepsilon<\infty$.

Use of Eq. (5.248) produces

$$\dot{V}_u(t)\leq-\lambda V_u(t)+\varepsilon, \tag{5.261}$$

where

$$\lambda=\frac{\lambda_3}{\lambda_2}. \tag{5.262}$$

Solution of Ineq. (5.261) produces

$$V_u(t)\leq V_u(0)\,e^{-\lambda t}+\frac{\varepsilon}{\lambda}. \tag{5.263}$$

Using Eq. (5.248) and the Ineq. (3.5) yields

$$\frac{\lambda_1\mathcal{B}v^2(x,t)}{L-a}\leq\lambda_1\mathcal{B}\int_a^L v_x^2dx\leq\lambda_1E_b^+\leq V_u(0)\,e^{-\lambda t}+\frac{\varepsilon}{\lambda}\in\mathcal{L}_\infty. \tag{5.264}$$

Thus $v(x,t)\in\mathcal{L}_\infty([a,L])$, so $u(x,t)\in\mathcal{L}_\infty([a,L])$.

□

Note that if we assume $v_t\in S([a,L])$, then the conditions of Lemma 53 imply $u_t\in\mathcal{L}_\infty([a,L])$.

Control Derivation

The EMK controller derived in this section exponentially regulates the acoustic displacement in the quiet subdomain but requires exact knowledge of the system parameters. To facilitate the control development, we first define the auxiliary variable

$$\eta(t) = \frac{1}{\mathcal{B}} p^-(t) + u_t(a,t). \tag{5.265}$$

where $p^-(t) = p(a^-,t)$. From Eqs. (2.39) and (5.265)

$$m\dot{\eta} = f - \mathbf{W}\Theta, \tag{5.266}$$

where the linear regression vector

$$\mathbf{W}(t) = [u_t(a,t), u(a,t), p_t^-(t) - p_t^+(t), p^-(t) - p^+(t), -p_t^-(t)], \tag{5.267}$$

and parameter vector

$$\Theta = \left[c, k, \frac{A\tau}{\mathcal{B}}, A, \frac{m}{\mathcal{B}}\right]^T. \tag{5.268}$$

We define $p_t^-(t) = p_t(a^-,t)$, $p_t^+(t) = p_t(a^+,t)$, and $p^+(t) = p(a^+,t)$.

Theorem 54 *The bounded control force*

$$f(t) = \mathbf{W}\Theta - k_p\eta \tag{5.269}$$

strongly exponentially stabilizes the acoustic displacement in the duct for $x \in [0,a]$.

Proof: Substitution of the EMK controller (5.269) into Eq. (5.266) yields

$$m\dot{\eta} = -k_p\eta. \tag{5.270}$$

To facilitate the subsequent stability analysis, we define the following non-negative scalar functions

$$E_b^-(t) = \alpha_1\left[\frac{\rho_0}{2}\int_0^a u_t^2 dx + \frac{\mathcal{B}}{2}\int_0^a u_x^2 dx\right], \tag{5.271}$$

$$E_c^-(t) = 2\alpha_2\rho_0\int_0^a x u_t u_x dx, \tag{5.272}$$

where α_1, $\alpha_2 > 0$.

We can bound $E_c^-(t)$ as follows

$$\lambda_{1e}E_b^- \le E_c^- \le \lambda_{2e}E_b^-,$$

where

$$\lambda_{11} = \left(1-\frac{2\alpha_2\rho_0 a}{\alpha_1 \min\left(\rho_0,\mathcal{B}\right)}\right),$$
$$\lambda_{21} = \left(1+\frac{2\alpha_2\rho_0 a}{\alpha_1 \min\left(\rho_0,\mathcal{B}\right)}\right),$$

and, for sufficiently small $\alpha_2 < \alpha_1$, $1 > \lambda_{11} > 0$ and $2 > \lambda_{21} > 1$.

We choose a Lyapunov functional

$$V(t) = E_b^- + E_c^- + \frac{1}{2}m\eta^2, \tag{5.273}$$

with

$$\lambda_{1e}\left(E_b^- + \eta^2\right) \leq V \leq \lambda_{2e}\left(E_b^- + \eta^2\right),$$

where

$$\lambda_{1e} = \min\left(\lambda_{11}, \frac{1}{2}m\right),$$
$$\lambda_{2e} = \max\left(\lambda_{21}, \frac{1}{2}m\right).$$

The time derivative of Eq. (5.273) depends on

$$\dot{E}_b^- = \alpha_1 \mathcal{B} u_x(a,t)\, u_t(a,t), \tag{5.274}$$
$$\dot{E}_c^- = 2\alpha_2\rho_0 \int_0^a x u_{tt} u_x dx + 2\alpha_2\rho_0 \int_0^a x u_t u_{xt} dx. \tag{5.275}$$

The first term in Eq. (5.275) simplifies as follows

$$\begin{aligned} A_1 &= 2\alpha_2\rho_0 \int_0^a x u_{tt} u_x dx \\ &= 2\alpha_2 \mathcal{B} \int_0^a x u_{xx} u_x dx \\ &= 2\alpha_2 \mathcal{B} x u_x^2 |_0^a - 2\alpha_2\mathcal{B} \int_0^a u_x^2 dx - A_1, \end{aligned}$$

yielding

$$A_1 = \alpha_2 \mathcal{B} a u_x^2(a,t) - \alpha_2 \mathcal{B} \int_0^a u_x^2 dx.$$

The second term in Eq. (5.275) simplifies in a similar manner to produce

$$\begin{aligned} A_2 &= 2\alpha_2\rho_0 \int_0^a x u_t u_{xt} dx \\ &= \alpha_2\rho_0 a_1 u_t^2(a,t) - \alpha_2\rho_0 \int_0^a u_t^2 dx. \end{aligned}$$

We now substitute Eqs. (5.274), (5.275), and (5.270) into

$$\begin{aligned}\dot{V}(t) &= \alpha_1 \mathcal{B} u_x(a,t)\, u_t(a,t) - k_p \eta^2 \\ &\quad + \alpha_2 \left(\rho_0 a u_t^2(a,t) + \mathcal{B} a u_x^2(a,t) - \frac{1}{2\alpha_1} E_b^- \right). \end{aligned} \tag{5.276}$$

Several terms in Eq. (5.276) combine as follows

$$\alpha_1 \mathcal{B} u_x(a,t)\, u_t(a,t) + \alpha_2 \rho_0 a u_t^2(a,t) + \alpha_2 \mathcal{B} a u_x^2(a,t) \le \delta_4 \eta^2,$$

where the small positive number

$$\delta_4 = \max(\alpha_1 \mathcal{B}, \alpha_2 \rho_0 a\,, \alpha_2 \mathcal{B} a) < k_p,$$

if α_1 and α_2 are sufficiently small.

Thus,

$$\begin{aligned}\dot{V} &\le -(k_p - \delta_4)\, \eta^2 - \frac{\alpha_2}{2\alpha_1} E_1 \\ &\le -\lambda_{3e} \left(E_b^- + \eta^2 \right) \\ &\le -\lambda_e V, \end{aligned}$$

and

$$V(t) \le V(0)\, e^{-\lambda_e t}, \tag{5.277}$$

where

$$\begin{aligned}\lambda_{3e} &= \min\left(k_p - \delta_4, \frac{\alpha_2}{2\alpha_1} \right), \\ \lambda_e &= \frac{\lambda_{3e}}{\lambda_{2e}}. \end{aligned}$$

From Eq. (3.16) we know that

$$\lambda_e \frac{\mathcal{B} u^2(x,t)}{a} \le \lambda_e \mathcal{B} \int_0^a u_x^2 dx \le \lambda_e E_b^- \le V(0)\, e^{-\lambda_e t}. \tag{5.278}$$

Thus for $x \in [0,a]$, $u(x,t) \to 0$ exponential fast.

We now ensure all signals remain bounded in closed loop operation. From Eq. (5.277), $V(t) \in \mathcal{L}_\infty$, so we know that $\int_0^a u_t^2 dx \in \mathcal{L}_\infty$. We assume $u_t(x,t) \in \mathcal{S}([0,a])$ so that boundedness of the kinetic energy implies $u_t(x,t) \in \mathcal{L}_\infty([0,a])$. From Ineq. (5.278) we have $u(x,t) \in \mathcal{L}_\infty([0,a])$. We assume $u_t(x,t) \in \mathcal{S}_{\mathcal{A}_0}([0,a])$ for an appropriately defined $\mathcal{A}_0$ so that $u_x(L,t) \in \mathcal{L}_\infty$ and $u_{xx}(L,t) \in \mathcal{L}_\infty([0,a])$. The field equation then implies that $u_{tt}(a,t) \in \mathcal{L}_\infty$. With $u(a,t)$, $u_t(a,t)$, and $u_{tt}(a,t) \in \mathcal{L}_\infty$, Lemma 53 implies that $u(x,t) \in \mathcal{L}_\infty([a,L])$. Thus, invoking similar assumptions for $x \in [a,L]$ yields p^+, $p_t^+ \in \mathcal{L}_\infty$, and all terms in the control law (5.269) are bounded, so $f(t) \in \mathcal{L}_\infty$.

□

5.2.2 Axially Moving Beam Isolator

Uncontrolled Subdomain Boundedness

In this example, we design a vibration isolating controller for the axially moving beam model given in Eqs. (2.58) - (2.63) . The control objective is to drive the beam displacement $w(x,t)$ for $x \in (0,a)$ to zero with a bounded control torque τ_c. We design an exponentially regulating control law for the system based on exact model knowledge. First, Lemma 55 shows that the uncontrolled subdomain displacement is bounded if $w_x(a,t)$ and its time derivatives and f_u are bounded.

Lemma 55 *If* $w_x(a,t)$, $w_{xt}(a,t)$, $w_{xtt}(a,t)$, *and* $f_u \in \mathcal{L}_\infty$, *then* $w(x,t) \in \mathcal{L}_\infty$, *for* $x \in [a,l]$

Proof: First, we transform Eq. (2.58) as follows

$$z(y,t) = w(l-y,t) - \frac{y^2}{l-a}\left(1-\frac{y}{l-a}\right) w_x(a,t) \text{ for } y \in [0,b], \tag{5.279}$$

where $y = l - x$ and $b = l - a$. Substitution of Eq. (5.279) into Eq. (2.58) yields

$$\rho z_{tt} - 2\rho v z_{yt} - P_0 z_{yy} + EI z_{yyyy} + c z_t = f(y,t), \quad y \in [0,b], \tag{5.280}$$

with boundary conditions:

$$z(0,t) = z_y(0,t) = 0, \tag{5.281}$$

$$z(b,t) = z_y(b,t) = 0, \tag{5.282}$$

where

$$\begin{aligned} f(y,t) &= f_u(l-y,t) + \frac{y^2}{l-a}\left(1-\frac{y}{l-a}\right)\rho w_{xtt}(a,t) - P_0\frac{2l-6y-2a}{(l-a)^2} w_x(a,t) \\ &+ \left(\frac{y^2}{l-a}\left(1-\frac{y}{l-a}\right)c - 2\rho v\left(\frac{2y}{l-a} - \frac{3y^2}{(l-a)^2}\right)\right) w_{xt}(a,t) \end{aligned} \tag{5.283}$$

is bounded by the Lemma assumptions.

A Lyapunov functional for the uncontrolled subdomain is chosen to be

$$V_u(t) = E_b^+(t) + E_c^+(t), \tag{5.284}$$

where

$$E_b^+ = \frac{1}{2}\rho\int_0^b z_t^2 dy + \frac{1}{2}P_0\int_0^b z_y^2 dy + \frac{1}{2}EI\int_0^b z_{yy}^2 dy, \tag{5.285}$$

$$E_c^+ = \beta\rho\int_0^b z z_t dy, \tag{5.286}$$

with $\beta > 0$ a small positive constant, and we assume that $P_0 + \frac{\pi^2}{4b^2}EI > 0$. Since $w(0) = 0$, we can use Ineq. (3.23) in Eq. (5.285) to produce

$$E_b^+ \geq \frac{1}{2}\rho \int_0^b z_t^2 dy + \frac{1}{2}\left(P_0 + \frac{\pi^2}{4b^2}EI\right)\int_0^b z_y^2 dy. \tag{5.287}$$

Substitution of Ineqs. (3.23) and (3.10) into Eq. (5.286) yields:

$$\begin{aligned} \left|E_c^+(t)\right| &\leq \beta\rho \int_0^b \left(z_t^2 + z^2\right) dy \leq \beta\rho \int_0^b \left(z_t^2 + \frac{4b^2}{\pi^2} z_y^2\right) dy \\ &\leq 2\beta\rho \frac{\max\left(1, \frac{4b^2}{\pi^2}\right)}{\min\left(\rho, P_0 + \frac{\pi^2}{4b^2}EI\right)} E_b^+(t). \end{aligned} \tag{5.288}$$

If we choose

$$\beta \leq \frac{\min\left(\rho, P_0 + \frac{\pi^2}{4b^2}EI\right)}{2\rho \max\left(1, \frac{4b^2}{\pi^2}\right)}$$

then

$$0 \leq \lambda_1 E_b^+(t) \leq V_u(t) \leq \lambda_2 E_b^+(t), \tag{5.289}$$

where

$$\lambda_1 = 1 - 2\beta\rho \frac{\max\left(1, \frac{4b^2}{\pi^2}\right)}{\min\left(\rho, P_0 + \frac{\pi^2}{4b^2}EI\right)} > 0, \tag{5.290}$$

$$\lambda_2 = 1 + 2\beta\rho \frac{\max\left(1, \frac{4b^2}{\pi^2}\right)}{\min\left(\rho, P_0 + \frac{\pi^2}{4b^2}EI\right)} > 1. \tag{5.291}$$

The time derivative of $V_u(t)$

$$\dot{V}_u(t) = \dot{E}_b^+(t) + \dot{E}_c^+(t) \tag{5.292}$$

depends on

$$\dot{E}_b^+ = -\int_0^b c z_t^2 dy + \int_0^b z_t f dy, \tag{5.293}$$

where the field equation (5.280), boundary conditions (5.281) and (5.282), and integration by parts have been applied. Application of Ineq. (3.12) produces

$$\dot{E}_b^+ \leq -(c - \delta_1) \int_0^b z_t^2 dy + \frac{1}{\delta_1} \int_0^b f^2 dy. \tag{5.294}$$

The remaining term in $\dot{V}_u(t)$ is

$$\begin{aligned} \dot{E}_c^+ &= \beta\rho \int_0^b \left(z_t^2 + z z_{tt}\right) dx \\ &= \beta \int_0^b \rho z_t^2 dy + \beta \int_0^b z\left(2\rho v z_{yt} + P_0 z_{yy} - EI z_{yyyy} - c z_t + f\right) dy. \end{aligned} \tag{5.295}$$

The terms in Eq. (5.295) simplify via integration by parts, the boundary conditions, and Ineqs. (3.23) and (3.12) as follows:

$$\begin{aligned}
2\beta\rho v\int_0^b zz_{yt}dy &\le \beta\rho v\int_0^b\left(\frac{1}{\delta_2}z_t^2+\delta_2 z_y^2\right)dy \\
&\le \beta\rho v\int_0^b\left(\frac{1}{\delta_2}z_t^2+\frac{4b^2}{\pi^2}\delta_2 z_{yy}^2\right)dy, \qquad (5.296)\\
\beta P_0\int_0^b zz_{yy}dy &= -\beta P_0\int_0^b z_y^2dy, \qquad (5.297)\\
-\beta EI\int_0^b zz_{yyyy}dy &= -\beta EI\int_0^b z_{yy}^2dy, \qquad (5.298)\\
-\beta c\int_0^b zz_tdy &\le \beta\frac{c}{\delta_3}\int_0^b z_t^2dy+\beta c\delta_3\frac{4b^2}{\pi^2}\int_0^b z_y^2dy \\
&\le \beta\frac{c}{\delta_3}\int_0^b z_t^2dy+\beta c\delta_3\frac{16b^4}{\pi^4}\int_0^b z_{yy}^2dy, \qquad (5.299)\\
\beta\int_0^b zfdy &\le \frac{\beta}{\delta_4}\int_0^b f^2dy+\beta\frac{4b^2}{\pi^2}\delta_4\int_0^b z_y^2dy \\
&\le \frac{\beta}{\delta_4}\int_0^b f^2dy+\beta\frac{16b^4}{\pi^4}\delta_4\int_0^b z_{yy}^2dy, \qquad (5.300)
\end{aligned}$$

where δ_2, δ_3, and δ_4 are arbitrary positive constants.

Substitution of Eqs. (5.296) through (5.300) into Eq. (5.292) yields:

$$\begin{aligned}
\dot{V}_u(t) &\le -\left(c-\delta_1-\beta\rho-\frac{\beta\rho v}{\delta_2}-\frac{\beta c}{\delta_3}\right)\int_0^b z_t^2dy \\
&\quad -\beta\left(EI-\rho v\frac{4b^2}{\pi^2}\delta_2-c\delta_3\frac{16b^4}{\pi^4}-\frac{16b^4}{\pi^4}\delta_4\right)\int_0^b z_{yy}^2dy \\
&\quad -\beta P_0\int_0^b z_y^2dy+\left(\frac{1}{\delta_1}+\frac{\beta}{\delta_4}\right)\int_0^b f^2dy \\
&\le -\lambda_3E_b^+(t)+\varepsilon, \qquad (5.301)
\end{aligned}$$

where

$$\lambda_3 = \min\left(\frac{\varepsilon_1}{\rho},\frac{\varepsilon_2}{EI},\beta\right)>0, \qquad (5.302)$$

$$\varepsilon = \left(\frac{1}{\delta_1}+\frac{\beta}{\delta_4}\right)\max\left(\int_0^b f^2dy\right)>0, \qquad (5.303)$$

$$\varepsilon_1 = c-\delta_1-\beta\rho-\frac{\beta\rho v}{\delta_2}-\frac{\beta c}{\delta_3}>0, \qquad (5.304)$$

$$\varepsilon_2 = \beta\left(EI-\rho v\frac{4b^2}{\pi^2}\delta_2-c\delta_3\frac{16b^4}{\pi^4}-\frac{16b^4}{\pi^4}\delta_4\right)>0, \qquad (5.305)$$

if β, δ_1, δ_2, δ_3, and δ_4 are sufficiently small, and $\varepsilon < \infty$ for all t.

Use of Eq. (5.289) produces

$$\dot{V}_u(t) \leq -\lambda V_u(t) + \varepsilon, \tag{5.306}$$

where $\lambda = \lambda_3/\lambda_2$, with the solution

$$V_u(t) \leq V_u(0) e^{-\lambda t} + \frac{\varepsilon}{\lambda} \in \mathcal{L}_\infty.$$

Using Ineq. (3.5) in Eq. (5.289) then yields,

$$\frac{\pi^2 EI}{8b^3}\lambda_1 z^2 \leq \frac{\pi^2 EI}{8b^2}\lambda_1 \int_0^b z_y^2 dy \leq \lambda_1 \frac{EI}{2}\int_0^b z_{yy}^2 dy \leq \lambda_1 E_b^+ \leq V_u(t) \in \mathcal{L}_\infty. \tag{5.307}$$

Thus $z \in \mathcal{L}_\infty$, and from Eq. (5.279), $w(x,t) \in \mathcal{L}_\infty([a,l])$.
□

Exact Model Knowledge Control Law

To facilitate the subsequent control design and analysis, we define an auxiliary variable,

$$\eta(t) = w_{xt}(a,t) + w_{xx}^-, \tag{5.308}$$

where $w_{xx}^- = w_{xx}(a^-,t)$. After differentiation of $\eta(t)$ with respect to time, multiplication by J, and substitution of the boundary condition (2.63), Eq. (5.308) becomes

$$J\dot{\eta}(t) = \tau_c + EI\left(w_{xx}^+ - w_{xx}^-\right) + Jw_{xxt}^-, \tag{5.309}$$

where $w_{xx}^+ = w_{xx}(a^+,t)$ and $w_{xxt}^- = w_{xxt}(a^-,t)$.

Theorem 56 *The bounded control torque,*

$$\tau_c = -EI\left(w_{xx}^+ - w_{xx}^-\right) - Jw_{xxt}^- - k_s\eta, \tag{5.310}$$

strongly exponentially stabilizes the beam displacement for $x \in [0,a]$.

Proof: Substitution of the EMK controller (5.310) into (5.309) yields

$$J\dot{\eta}(t) = -k_s\eta. \tag{5.311}$$

A Lyapunov functional is defined

$$V_c(t) = E_b^-(t) + E_c^-(t) + \frac{1}{2}J\eta^2, \tag{5.312}$$

where

$$E_b^- = \alpha\left(\frac{1}{2}\rho\int_0^a w_t^2 dx + \frac{1}{2}P_0\int_0^a w_x^2 dx + \frac{1}{2}EI\int_0^a w_{xx}^2 dx\right), \quad (5.313)$$

$$E_c^- = 2\beta\rho\int_0^a \gamma(x)\, w_x\,(w_t + v w_x)\, dx, \quad (5.314)$$

$\gamma(x) = x - a$, and α and β are small positive constants. As in Eq. (5.287), we know

$$E_b^- \geq \frac{1}{2}\rho\int_0^a w_t^2 dx + \frac{1}{2}\left(P_0 + \frac{\pi^2}{4a^2}EI\right)\int_0^a w_x^2 dx. \quad (5.315)$$

Integration by parts bounds Eq. (5.314) as follows:

$$\begin{aligned}\left|E_c^-\right| &\leq 2\beta\rho a\int_0^a \left(\left|w_x w_t\right| + \left|v w_x^2\right|\right) dx \leq 2\beta\rho a\int_0^a \left(w_t^2 + (1+v)\, w_x^2\right) dx \\ &\leq 4\beta\rho a\,(1+v)\left[\frac{1}{2}\int_0^a \left(w_t^2 + w_x^2\right) dx\right] \\ &\leq \frac{4\beta\rho a\,(1+v)}{\min\left(\rho, P_0 + \frac{\pi^2}{4a^2}EI\right)\alpha}\left[\frac{1}{2}\alpha\int_0^a \left(\rho w_t^2 + P_0 w_x^2 + EI w_{xx}^2\right) dx\right], \quad (5.316)\end{aligned}$$

so

$$-\xi E_b^- \leq E_c^- \leq \xi E_b^-, \quad (5.317)$$

where

$$\xi = \frac{4\beta\rho a\,(1+v)}{\min\left(\rho,\ P_0 + \frac{\pi^2}{4a^2}EI\right)\alpha} < 1, \quad (5.318)$$

if β is sufficient small. Thus

$$\lambda_{1e}\left(E_b^- + \eta^2\right) \leq V_c(t) \leq \lambda_{2e}\left(E_b^- + \eta^2\right), \quad (5.319)$$

where

$$\lambda_{1e} = \min\left(1 - \xi,\ \frac{J}{2}\right) > 0, \quad (5.320)$$

$$\lambda_{2e} = \max\left(1 + \xi,\ \frac{J}{2}\right) > 1. \quad (5.321)$$

The time derivative of Eq. (5.312) is

$$\dot{V}_c(t) = \dot{E}_b^- + \dot{E}_c^- - k_s\eta^2, \quad (5.322)$$

where

$$\frac{1}{\alpha}\dot{E}_b^- = EI w_{xt}(a,t)\, w_{xx}^- \leq -\frac{EI}{2}\left(w_{xt}^2(a,t) + \left(w_{xx}^-\right)^2\right) + \frac{EI}{2}\eta^2, \quad (5.323)$$

$$\begin{aligned}\frac{1}{\beta}\dot{E}_c^- &= 2\rho\int_0^a \gamma(x)\, w_{xt} w_t dx + 2\rho\int_0^a \gamma(x)\, w_x w_{tt} dx \\ &\quad + 4\rho v\int_0^a \gamma(x)\, w_x w_{xt} dx. \quad (5.324)\end{aligned}$$

Simplification of the terms in Eq. (5.324) using Ineq. (3.12) produces

$$2\rho \int_0^a \gamma(x)\, w_{xt} w_t dx = -\rho \int_0^a w_t^2 dx, \tag{5.325}$$

$$2P_0 \int_0^a \gamma(x)\, w_x w_{xx} dx = -P_0 \int_0^a w_x^2 dx, \tag{5.326}$$

$$\begin{aligned} -2EI \int_0^a \gamma(x)\, w_x w_{xxxx} dx &\leq -(3-\delta_5 a)\, EI \int_0^a w_{xx}^2 dx \\ &\quad + EI \frac{1}{\delta_5} w_{xx}^2(a), \end{aligned} \tag{5.327}$$

where, for simplicity, we neglect the stabilizing effect of damping in the controlled subdomain. Thus,

$$\begin{aligned} \dot{V}_c(t) &= -\beta\rho \int_0^a w_t^2 dx - \beta P_0 \int_0^a w_x^2 dx - \beta(3-\delta_5 a)\, EI \int_0^{a^-} w_{xx}^2 dx \\ &\quad - \left(k_s - \frac{EI}{2}\alpha\right)\eta^2 - \frac{EI}{2}\alpha w_{xt}^2(a,t) \\ &\quad - \left(\frac{EI}{2}\alpha - EI\frac{\beta}{\delta_5}\right)\left(w_{xx}^-\right)^2. \end{aligned} \tag{5.328}$$

If α, β, and δ_5 are sufficient small,

$$\dot{V}_c(t) \leq -\lambda_{3e} E_b^- \leq -\frac{\lambda_{3e}}{\lambda_{2e}} V_c(t), \tag{5.329}$$

where

$$\lambda_{3e} = \min\left(\frac{2\beta}{\alpha}, \frac{2\beta(3-\delta_5 a)}{\alpha}, \frac{2k_s - EI\alpha}{J}\right) > 0, \tag{5.330}$$

if $\alpha < 2k_s/EI$, and $\delta_5 < 3/a$. Thus

$$V_c(t) \leq V_c(0) \exp\left(-\frac{\lambda_{3e}}{\lambda_{2e}} t\right), \tag{5.331}$$

and using Ineq. (3.23),

$$\begin{aligned} w^2(x,t) &\leq a \int_0^a w_x^2 dx \leq \frac{8a^3}{\alpha\pi^2 EI} \frac{EI}{2}\alpha \int_0^a w_{xx}^2 dx \\ &\leq \frac{8a^3}{\alpha\pi^2 EI} E_b^- \leq \frac{8a^3}{\alpha\pi^2 EI \lambda_{1e}} V_c(t), \end{aligned} \tag{5.332}$$

so $w(x,t) \to 0$ as $t \to \infty$ exponentially fast.

We assume that $w(x,t) \in \mathcal{S}_{A_0}([0,a])$ so $w(x,t) \in \mathcal{L}_\infty([0,a])$ implies $w_x(x,t)$ and $w_{xx}(x,t) \in \mathcal{L}_\infty([0,a])$. The boundedness of $V_c(t)$ implies $\int_0^a w_t^2 dx \in \mathcal{L}_\infty$ and we

assume that $w_t(x,t) \in S([0,a])$ so $w_t(x,t) \in \mathcal{L}_\infty([0,a])$. Finally, we assume $w(x,t) \in S_{\mathcal{A}_0}([0,a])$ so w_{xt} and $w_{xxt} \in \mathcal{L}_\infty([0,a])$. From Eqs. (5.308), (5.312), and (5.331), $\eta \in \mathcal{L}_\infty$, so from Eq. (5.311) $\dot{\eta} \in \mathcal{L}_\infty$. Thus, $w_{xtt}(a,t) \in \mathcal{L}_\infty$, and from Lemma 55, $w_{xx}^+ \in \mathcal{L}_\infty$. Thus, all terms in the exact model knowledge control law are bounded so $\tau_c \in \mathcal{L}_\infty$.
□

We may redefine

$$\eta(t) = w_{xt}(a,t) + \alpha_1 w_{xx}^- \tag{5.333}$$

to allow fine tuning of the controller using additional gain α_1 without changing the closed loop stability.

It is interesting to note that the EMK control law (5.310) does not depend on the material tension because the beam is pinned at the actuator location. Friction at the actuator could raise the tension in the uncontrolled subdomain relative to the controlled subdomain. Substitution of the upstream P_0^- and downstream P_0^+ tension in the field equation (2.58) and functionals (5.285) and (5.313) are the only changes needed to incorporate unequal, constant tensions in the two subdomains.

5.3 Modal Control using Distributed Sensing

In this section, we develop a modal controller for conservative flexible systems that avoids spillover instabilities by employing distributed sensing. The control approach can be used with modal sensors [70] or with distributed sensing from high speed video. The controller includes a general spatial filter design based on the system eigenfunctions. Observer-based feedback exponentially stabilizes N controlled modes. A Lyapunov-like stability analysis proves that the residual modes remain bounded in closed-loop operation. We experimentally implement a single mode controller for a flexible link robot using high-speed video feedback and demonstrate the closed-loop performance.

5.3.1 *Mathematical Model*

Neglect of damping and gyroscopic effects in the matrix operator Eq. (2.86) produces

$$\ddot{\mathbf{w}} + \mathcal{A}_0\mathbf{w} = \mathcal{B}_0\mathbf{f}. \tag{5.334}$$

We assume that operator $\mathcal{A}_0$ is symmetric and positive definite as follows

$$(\mathbf{w}, \mathcal{A}_0\mathbf{w}) > \gamma \|\mathbf{w}\|^2 \text{ for all } \mathbf{w} \in \mathcal{H},\ \mathbf{w} \neq 0, \tag{5.335}$$

where γ is a positive scalar constant and the operator $\mathcal{B}_0$ is bounded in the sense that

$$\|\mathcal{B}_0\mathbf{f}\| \leq \alpha \|\mathbf{f}\|_2, \tag{5.336}$$

where α is a positive scalar constant.

To obtain the modal equations of motion, we assume the existence of a separable solution for Eq. (5.334) of the form given in Eq. (2.118) producing the eigenvalue problem in Eq. (2.119). Under the assumption of distinct ω_i, the eigenfunctions are orthogonal according to Eqs. (2.120) and (2.121).

The displacement is transformed into the first N modes and a residual displacement $\mathbf{v}(\mathbf{x},t)$ as follows:

$$\mathbf{w} = \sum_{k=1}^{N} \mathbf{W}_k \eta_k(t) + \mathbf{v}(\mathbf{x},t), \tag{5.337}$$

where the subscript k denotes the kth controlled mode, $\eta_k(t)$ denotes the kth modal coordinate, and $\mathbf{v}(\mathbf{x},t)$ is orthogonal to the first N controlled modes

$$(\mathbf{W}_k, \mathbf{v}) = (\mathcal{A}_0 \mathbf{W}_k, \mathbf{v}) = 0 \quad \forall k \leq N. \tag{5.338}$$

This implies that

$$(\mathbf{v}, \mathcal{A}_0 \mathbf{w}) = (\mathbf{v}, \mathcal{A}_0 \mathbf{v}) \tag{5.339}$$

and

$$(\mathbf{v}, \mathbf{w}) = (\mathbf{v}, \mathbf{v}). \tag{5.340}$$

Separability means

$$(\dot{\mathbf{v}}, \mathcal{A}_0 \mathbf{w}) = (\dot{\mathbf{v}}, \mathcal{A}_0 \mathbf{v}) \tag{5.341}$$

and

$$(\dot{\mathbf{v}}, \ddot{\mathbf{w}}) = (\dot{\mathbf{v}}, \ddot{\mathbf{v}}). \tag{5.342}$$

To obtain the modal equations of motion, we evaluate the inner product

$$(\mathbf{W}_k, \ddot{\mathbf{w}} + \mathcal{A}_0 \mathbf{w} = \mathcal{B}_0 \mathbf{f}\ ) \quad \forall k = 1, .., N. \tag{5.343}$$

Substitution of Eq. (5.337) into Eq. (5.343) and use of the orthogonality relations in Eqs. (2.120), (2.121), and (5.338) produce the N linearly independent equations

$$\ddot{\eta}_k(t) + \omega_k^2 \eta_k(t) = \mathbf{b}_k^T \mathbf{f} \qquad \forall k = 1, 2, ..., N, \tag{5.344}$$

where $\mathbf{b}_k^T \mathbf{f} = (\mathbf{W}_k, \mathcal{B}_0 \mathbf{f})$ and we assume that $\|\mathbf{b}_k\|_2 \neq 0\ \forall k \leq N$ so that the controlled modes are controllable. These modal equations of motion form the open-loop system

$$\ddot{\boldsymbol{\eta}} + \boldsymbol{\Omega}_N^2 \boldsymbol{\eta} = \mathbf{B}_0 \mathbf{f}, \tag{5.345}$$

where $\boldsymbol{\Omega}_N^2 = \text{diag}\,\{\omega_k^2\}$, $\mathbf{B}_0 = [\mathbf{b}_1, \mathbf{b}_2, ..., \mathbf{b}_N]^T$, and $\boldsymbol{\eta} = [\eta_1, \eta_2, ..., \eta_N]^T$.

5.3.2 Control Design

The control objective is to exponentially regulate the controlled modes in Eq. (5.345) without destabilizing the residual Eq. (5.359). We assume that the distributed measurements $\mathbf{w}(\mathbf{x},t)$ are available for all $\mathbf{x}$ and t. A full-state observer is constructed to estimate the modal velocities from the $\mathbf{w}(\mathbf{x}, t)$ data. The modal coordinates are calculated from the mode shapes and distributed measurements as follows:

$$\eta_k = (\mathbf{W}_k, \mathbf{w}). \tag{5.346}$$

The modal equations (5.345) are written in state-space form

$$\begin{aligned} \dot{\mathbf{x}} &= \mathbf{A}\mathbf{x} + \mathbf{B}\mathbf{f} \\ \mathbf{y} &= \mathbf{C}\mathbf{x}, \end{aligned} \tag{5.347}$$

where $\mathbf{y} = \boldsymbol{\eta}$ is the output, the state vector $\mathbf{x} = \begin{bmatrix} \boldsymbol{\eta}^T & \dot{\boldsymbol{\eta}}^T \end{bmatrix}^T \in \mathbb{R}^{2N}$ and the matrices

$$\mathbf{A} = \begin{bmatrix} \mathbf{0} & \mathbf{I} \\ -\boldsymbol{\Omega}_N^2 & \mathbf{0} \end{bmatrix}, \quad \mathbf{B} = \begin{bmatrix} \mathbf{0} \\ \mathbf{B}_0 \end{bmatrix}, \text{ and } \mathbf{C} = \begin{bmatrix} \mathbf{I} & \mathbf{0} \end{bmatrix}. \tag{5.348}$$

A state observer

$$\dot{\hat{\mathbf{x}}} = \mathbf{A}\hat{\mathbf{x}} + \mathbf{B}\mathbf{f} + \mathbf{L}(\mathbf{y} - \mathbf{C}\hat{\mathbf{x}}) \tag{5.349}$$

produces the state estimate $\hat{\mathbf{x}}(t)$. The gain matrix $\mathbf{L}$ is used to tune the observer time response. The observation error is defined as

$$\tilde{\mathbf{x}}(t) = \mathbf{x}(t) - \hat{\mathbf{x}}(t) \tag{5.350}$$

and the observation error dynamics are

$$\dot{\tilde{\mathbf{x}}} = (\mathbf{A} - \mathbf{L}\mathbf{C})\,\tilde{\mathbf{x}}. \tag{5.351}$$

Observed state feedback is designed as follows:

$$\mathbf{f} = -\mathbf{K}_s\hat{\mathbf{x}}, \tag{5.352}$$

where $\mathbf{K}_s$ is the feedback gain matrix. With the control law of Eq. (5.352), the state equation becomes

$$\dot{\mathbf{x}} = (\mathbf{A} - \mathbf{B}\mathbf{K}_s)\,\mathbf{x} + \mathbf{B}\mathbf{K}_s\tilde{\mathbf{x}}. \tag{5.353}$$

From Eqs. (5.351) and (5.353), we formulate the dynamics of the observed-state feedback control system

$$\begin{bmatrix} \dot{\mathbf{x}} \\ \dot{\tilde{\mathbf{x}}} \end{bmatrix} = \begin{bmatrix} \mathbf{A} - \mathbf{B}\mathbf{K}_s & \mathbf{B}\mathbf{K}_s \\ \mathbf{0} & \mathbf{A} - \mathbf{L}\mathbf{C} \end{bmatrix} \begin{bmatrix} \mathbf{x} \\ \tilde{\mathbf{x}} \end{bmatrix}, \tag{5.354}$$

with the characteristic equation

$$|s\mathbf{I} - \mathbf{A} + \mathbf{B}\mathbf{K}_s|\,|s\mathbf{I} - \mathbf{A} + \mathbf{L}\mathbf{C}| = 0. \tag{5.355}$$

The matrices $\mathbf{L}$ and $\mathbf{K}_s$ are designed to ensure negative closed-loop roots of the characteristic Eq. (5.355) such that $\mathbf{x}$ and $\tilde{\mathbf{x}}$ in Eq. (5.354) exponentially decay to zero. Hence, from Eq. (5.350), $\hat{\mathbf{x}}$ is exponentially stable in the sense that

$$\|\hat{\mathbf{x}}\|_2 \leq \xi_1 \exp(-\xi_2 t), \tag{5.356}$$

where ξ_1 and ξ_2 are positive scalar constants. Substitution of the bound of Eq. (5.356) into Eq. (5.352) yields

$$\|\mathbf{f}\|_2 = \|[\mathbf{K}_s]\,\hat{\mathbf{x}}\|_2 \leq \beta\xi_1 \exp(-\xi_2 t), \tag{5.357}$$

where $\beta = \|\mathbf{K}_s\|_2$.

5.3.3 *Residual Mode Stability*

The weak form of the residual equations of motion results from the inner product

$$(\dot{\mathbf{v}}, \ddot{\mathbf{w}} + \mathcal{A}_0\mathbf{w} = \mathcal{B}_0\mathbf{f}) \tag{5.358}$$

producing

$$(\dot{\mathbf{v}}, \ddot{\mathbf{v}}) = -(\dot{\mathbf{v}}, \mathcal{A}_0\mathbf{v}) + (\dot{\mathbf{v}}, \mathcal{B}_0\mathbf{f}), \tag{5.359}$$

where the orthogonality property in Eq. (5.338) has been used.

To establish the residual mode stability result, we formulate the energy of the system as

$$E_r = \frac{1}{2}\|\dot{\mathbf{v}}\|^2 + \frac{1}{2}(\mathbf{v}, \mathcal{A}_0\mathbf{v}), \tag{5.360}$$

where the first and second terms in Eq. (5.360) represent the kinetic and potential energies, respectively. The time derivative of Eq. (5.360) yields

$$\dot{E}_r = (\dot{\mathbf{v}}, \ddot{\mathbf{v}}) + (\dot{\mathbf{v}}, \mathcal{A}_0\mathbf{v}) = (\dot{\mathbf{v}}, \mathcal{B}\mathbf{f}), \tag{5.361}$$

where Eq. (5.359) and the distributive property of the inner product is used. Eq. (5.361) is upper bounded as

$$\dot{E}_r \leq |(\dot{\mathbf{v}}, \mathcal{B}_0\mathbf{f})| \leq \|\dot{\mathbf{v}}\|\,\|\mathcal{B}_0\mathbf{f}\|, \tag{5.362}$$

where the Cauchy-Schwarz Ineq. (3.4) is used. Substitution of the bounds in Ineqs. (5.336) and (5.357) into Eq. (5.359) produces

$$\|\mathcal{B}_0\mathbf{f}\| \leq \alpha\beta\xi_1 \exp(-\xi_2 t). \tag{5.363}$$

After substituting Ineq. (5.363) into Ineq. (5.362), we obtain

$$\dot{E}_r \leq \sqrt{\|\dot{\mathbf{v}}\|^2}\alpha\beta\xi_1 \exp(-\xi_2 t) \leq \sqrt{\|\dot{\mathbf{v}}\|^2 + \|\mathbf{v}\|^2}\alpha\beta\xi_1 \exp(-\xi_2 t). \tag{5.364}$$

Use of Ineq. (5.335) in Ineq. (5.364) produces

$$\dot{E}_r \leq \left(\sqrt{(\dot{\mathbf{v}}, \dot{\mathbf{v}}) + \frac{1}{\gamma} (\mathbf{v}, \mathcal{A}_0 \mathbf{v})} \right) \alpha \beta \xi_1 \exp\left(-\xi_2 t\right). \tag{5.365}$$

After substituting Eq. (5.360) into Ineq. (5.365), we obtain the upper bound

$$\dot{E}_r \leq \max\left\{ \sqrt{2}, \sqrt{\frac{2}{\gamma}} \right\} \sqrt{E_r} \alpha \beta \xi_1 \exp\left(-\xi_2 t\right), \tag{5.366}$$

which is rewritten as

$$\frac{d}{dt}\left(2\sqrt{E_r}\right) \leq \max\left\{ \sqrt{2}, \sqrt{\frac{2}{\gamma}} \right\} \alpha \beta \xi_1 \exp\left(-\xi_2 t\right). \tag{5.367}$$

The integration of Ineq. (5.367) yields

$$\sqrt{E_r} \leq \xi_3 \left[1 - \exp\left(-\xi_2 t\right)\right] + c_2, \tag{5.368}$$

where c_2 is the constant of integration and the positive scalar constant

$$\xi_3 = \frac{1}{2} \max\left\{ \sqrt{2}, \sqrt{\frac{2}{\gamma}} \right\} \frac{\alpha \beta \xi_1}{\xi_2}. \tag{5.369}$$

Ineq. (5.368) is rewritten as

$$\sqrt{E_r} \leq \xi_3 \exp\left(-\xi_2 t\right) + c_1, \tag{5.370}$$

where the positive scalar constant $c_1 = \xi_3 + c_2$. Upon squaring both sides of Ineq. (5.370), we obtain

$$E_r \leq \gamma_3 \exp\left(-\gamma_4 t\right) + c_0, \tag{5.371}$$

where positive scalar constants

$$\gamma_3 = \max\left\{\xi_3^2, 2\xi_3 c_1\right\}, \; \gamma_4 = \min\left\{\xi_2, 2\xi_2\right\}, \;\; c_0 = c_1^2. \tag{5.372}$$

Thus, from Eq. (5.360) and Ineqs. (5.371) and (5.335), we obtain

$$\|\mathbf{v}\|^2 \leq \frac{1}{2\gamma}\left(\gamma_3 \exp\left(-\gamma_4 t\right) + c_0\right) \tag{5.373}$$

and the residual modes remain bounded in closed-loop operation.

5.3.4 Flexible Link Robot Arm

System Model

The flexible link robot arm operator equations are given by Eq. (2.110). Using the inner product defined in Eq. (2.111) and the defined Hilbert space, we see that $\mathcal{A}_0$ is symmetric. Clearly, $\mathcal{B}_0$ satisfies Eq. (5.336) with $\alpha = 1$. The following Lemma shows that $\mathcal{A}_0$ satisfies Eq. (5.335).

Lemma 57 *The operator $\mathcal{A}_0$ given in Eq. (2.110) is positive-definite as indicated by Eq. (5.335) with* $\gamma = \min\left\{\frac{EI}{4\rho L^4}, \frac{EI}{4mL^3}, \frac{EI}{4J_m L}, \frac{EI}{4J_h L}\right\}$.

Proof:

Based on the inner product definition of Eq. (2.111),

$$\begin{aligned}(\mathbf{w}, \mathcal{A}_0\mathbf{w}) &= EI\left[\int_0^L w(x)\, w_{xxxx}(x)\, dx - w(L)\, w_{xxx}(L)\right. \\ &\quad \left. + w_x(L)\, w_{xx}(L) - w_x(0)\, w_{xx}(0)\right]. \end{aligned} \tag{5.374}$$

Integrating by parts the first term on the right-hand side of Eq. (5.374) twice, we obtain

$$\begin{aligned}(\mathbf{w}, \mathcal{A}_0\mathbf{w}) &= EI\left[\int_0^L w_{xx}(x)\, w_{xx}(x)\, dx + w w_{xxx}\big|_0^L - w_x w_{xx}\big|_0^L\right. \\ &\quad \left. - w(L)\, w_{xxx}(L) + w_x(L)\, w_{xx}(L) - w_x(0)\, u_{xx}(0)\right]. \end{aligned} \tag{5.375}$$

Cancelling common terms and applying the boundary condition of Eq. (2.54), we obtain

$$(\mathbf{w}, \mathcal{A}_0\mathbf{w}) = EI\int_0^L w_{xx}^2(x)\, dx, \tag{5.376}$$

which is rewritten as

$$\begin{aligned}(\mathbf{w}, \mathcal{A}_0\mathbf{w}) &= \frac{EI}{4}\int_0^L w_{xx}^2(x)\, dx + \frac{EI}{4}\int_0^L w_{xx}^2(x)\, dx \\ &\quad + \frac{EI}{4}\int_0^L w_{xx}^2(x)\, dx + \frac{EI}{4}\int_0^L w_{xx}^2(x)\, dx. \end{aligned} \tag{5.377}$$

After applying Ineqs. (3.24), (3.5), and (3.9) to Eq. (5.377), we obtain

$$(\mathbf{w}, \mathcal{A}_0\mathbf{w}) \geq \frac{EI}{4L^4}\int_0^L w^2(x) + \frac{EI}{4L^3}w^2(x,t) + \frac{EI}{4L}w_x^2(x,t) + \frac{EI}{4L}w_x^2(x,t)\, dx. \tag{5.378}$$

The last three terms on the right hand side of Ineq. (5.378) are pointwise quantities (*i.e.*, $\forall x \in [0, L]$). Hence, it follows that

$$\begin{aligned} w^2(x,t) &\geq w^2(L,t), \quad w_x^2(x,t) \geq w_x^2(L,t), \\ w_x^2(x,t) &\geq w_x^2(0,t) \quad \forall x \in [0, L]. \end{aligned} \tag{5.379}$$

After substituting Ineqs. (5.379) into Ineq. (5.378), we obtain

$$(\mathbf{w}, \mathcal{A}_0\mathbf{w}) \geq \min\left\{\tfrac{EI}{4\rho L^4}, \tfrac{EI}{4mL^3}, \tfrac{EI}{4J_mL}, \tfrac{EI}{4J_hL}\right\}\left[\rho\int_0^L w^2(x)\,dx + mw^2(L,t) + J_m w_x^2(L,t) + J_h w_x^2(0,t)\right], \tag{5.380}$$

which is rewritten by using Eq. (2.111) as

$$(\mathbf{w}, \mathcal{A}_0\mathbf{w}) \geq \min\left\{\frac{EI}{4\rho L^4}, \frac{EI}{4mL^3}, \frac{EI}{4J_mL}, \frac{EI}{4J_hL}\right\}\|\mathbf{w}\|^2 \triangleq \gamma\|\mathbf{w}\|^2 \tag{5.381}$$

and is of the form given in Ineq. (5.335).
□

Experimental Validation

The experimental testbed shown in Figure 10 consists of thin flexible beam actuated at one end by a switched reluctance motor (SRM) and carrying a payload mass of 0.1 [kg]. The following control hardware is used: (i) a Dalsa CAD-6 camera that captures 955 frames per second with 8-bit gray scale at a 256×256 pixel resolution; (ii) a Road Runner Model 24 video capture board; and (iii) two Pentium II-based personal computers (PCs) operating under QNX (micro kernel-based, real-time operating system). A 102,400-count resolver is mounted on the SRM to measure hub angular displacement $w_x(0,t)$. Data acquisition and control implementation are performed at 1 [kHz] via the Quanser MultiQ I/O Board and interfacing circuitry. The Dalsa camera, with lens of 0.08 [m] focal length, is mounted 1.2 [m] above the robot workspace. One PC hosts the video capture board and acquires and processes the visual data from the high-speed camera. This raw data is processed to obtain distributed displacement measurements. Figure 11 (a) shows a photograph of the experimental setup as seen by the camera. The camera viewing area is square and, to simplify the image processing, it does not include the entire link. A snapshot, the beam centerline pixel data, and a best fit cubic polynomial are shown in Figures 11 (b) – (d). The four time-varying polynomial coefficients are transmitted via a fast, dedicated TCP/IP connection to the control PC, where the control algorithm and other I/O operations associated with the flexible link robot are implemented.

The various parameters associated with the flexible link robot system are calculated using standard test procedures and engineering handbook tables to be, $m = 0.31$ [kg], $L = 1$ [m], $EI = 2.933$ [N–m^2], $J_h = 0.034$ [kg–m^2], $J_m = 0.1562$ [kg–m^2], $\rho = 0.239$ [kg/m].

The first modal coordinate η_1 is numerically computed on-line according to Eq. (5.346). The time derivative calculations are implemented using a standard, backwards difference/filtering algorithm for the full-state feedback implementation only. The control algorithms are written in C++ and implemented using the QMotor [104] real-time control environment.

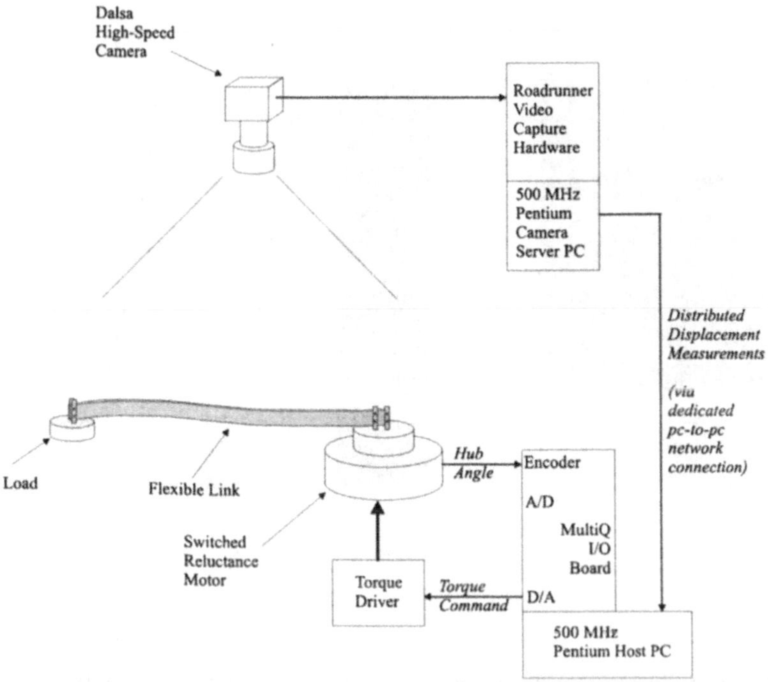

Figure 10 Schematic diagram of the experimental setup for the flexible link robot.

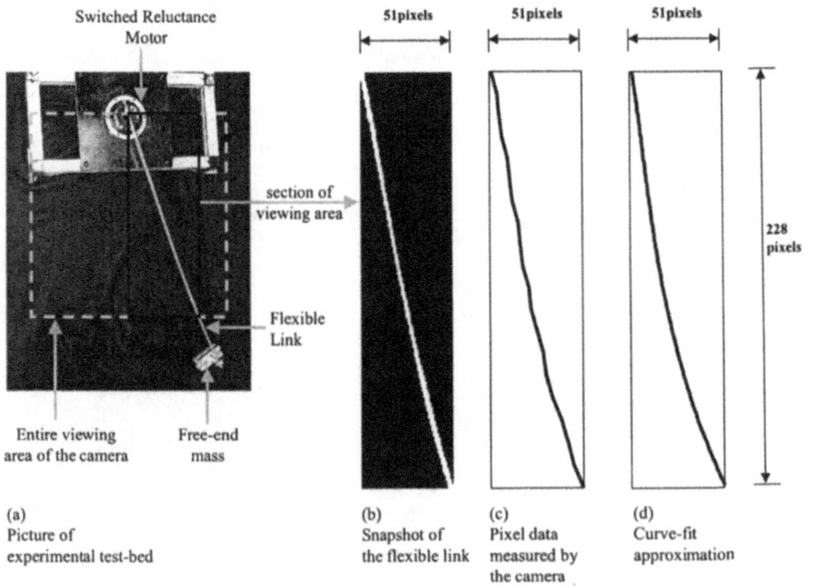

Figure 11 Experimental setup and image processing: (a) Photograph of the experimental setup showing camera viewing area, (b) Snapshot of the beam taken by the camera, (c) Pixel data measured by the camera, and (d) Third order polynomial curve-fit approximation of the pixel data.

For the flexible link robot experiments, we choose $N = 1$ to focus control on the first flexible mode. Implementation of the proposed controller in Eq. (5.352) requires measurement of $\eta_1(t)$ as shown in Eq. (5.346). Consequently, determination of the mode shape W_1 is essential. Solving for the lowest frequency root of Eq. (2.153) using experimental parameters yields

$$\beta_1 = 1.483. \tag{5.382}$$

The constant coefficients in Eq. (2.152) are numerically computed to yield the first eigenfunction

$$\begin{aligned} \mathbf{W}_1(x) &= -0.9546\sin(\beta_1 x) - 0.1959\cos(\beta_1 x) \\ &\quad +0.1098\sinh(\beta_1 x) + 0.1959\cosh(\beta_1 x). \end{aligned} \tag{5.383}$$

The eigenvalue in Eq. (5.382) corresponds to a natural frequency of 1.23 [Hz] for the first flexible mode. This analytical result is experimentally validated to be 1.22 [Hz] by energizing the robot actuator with a small sinusoidal torque trajectory with a time-varying frequency from 0.1 [Hz] to 4 [Hz] over an interval of 10 [sec] and extracting the frequency of peak link angular displacement.

The objective of the experiment is to regulate the angular displacement of the flexible link robot to a setpoint of 20 [deg] starting from an initial angle of 0 [deg].

Experiment 1: For comparison, a PD control law

$$\tau_h(t) = -k_a w_x(0,t) - k_b w_{xt}(0,t),$$

is implemented with $k_a = 2.0$ and $k_b = 0.01$ using the hub encoder feedback. The control gains are tuned to achieve the best performance. The plots for the PD control implementation are shown in Figure 12.

Experiment 2: A full-state feedback version of the proposed controller defined in Eq. (5.352) is implemented with

$$\mathbf{f} = -\bar{k}_d\dot{\eta}_1 - \bar{k}_p\eta_1, \tag{5.384}$$

where

$$\begin{aligned} \eta_1 &= \rho\int_0^L W_1(x)\,w\,dx + J_h W_{1x}(0)\,w_x(0,t) \\ &\quad +mW_1(L)\,w(L,t) + J_m W_{1x}(L)\,w_x(L,t) \end{aligned} \tag{5.385}$$

and $\dot{\eta}_1$ is obtained by backwards difference/filtering of η_1. The best performing control gains are

$$\bar{k}_p = 3.0 \qquad \bar{k}_d = 0.03$$

and the plots for the full-state feedback implementation are shown in Figure 13.

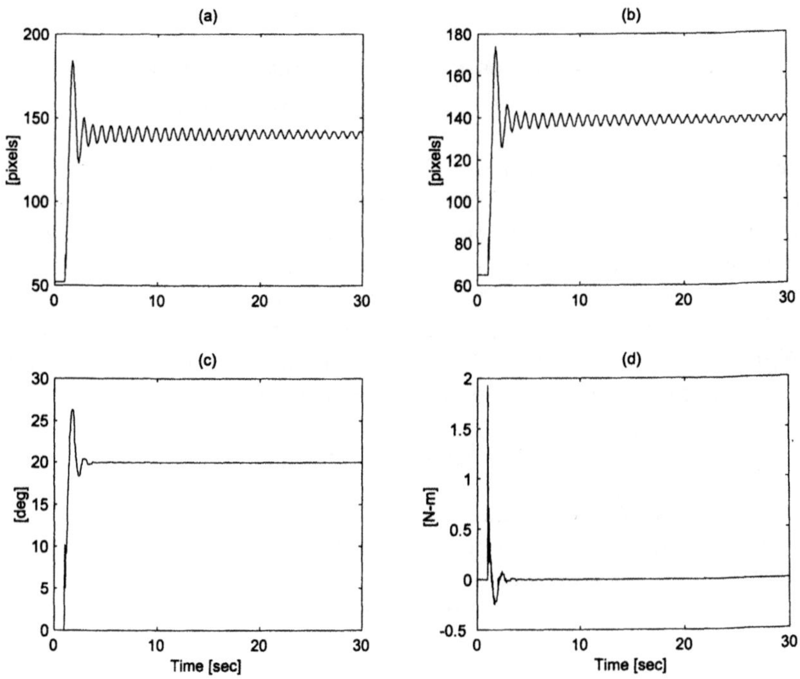

Figure 12 Experimental PD control: (a) Beam displacement at $x = 0.45L$, (b) Beam displacement at $x = 0.85L$, (c) Hub angle, and (d) Applied hub torque.

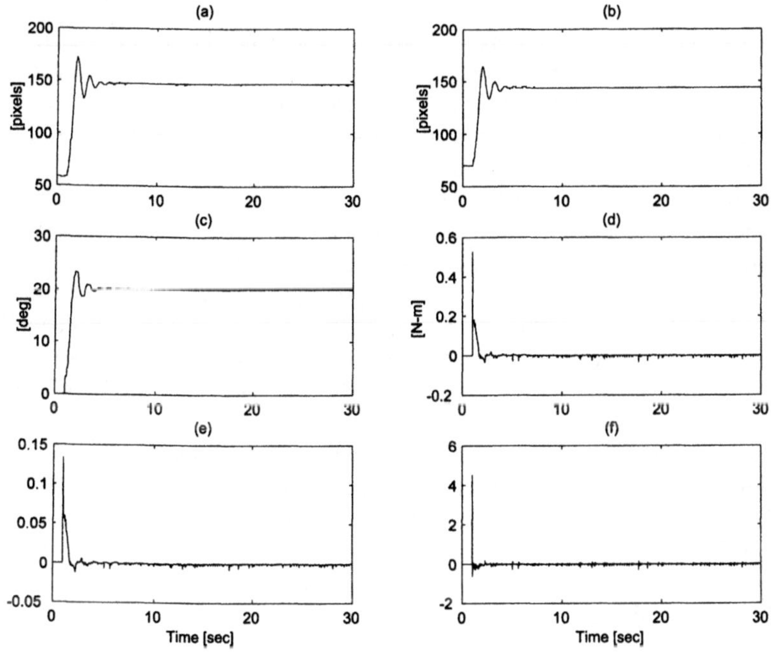

Figure 13 Experimental full order modal control: (a) Beam displacement at $x = 0.45L$, (b) Beam displacement at $x = 0.85L$, (c) Hub angle, (d) Applied hub torque, (e) Modal coordinate η, and (f) Time derivative of modal coordinate $\dot{\eta}$.

Experiment 3: The proposed controller defined by Eq. (5.352) is implemented, where $\dot{\hat{\eta}}_1$is obtained by integrating the equation

$$\ddot{\hat{\eta}}_1 = \ -\left(1+\omega_1^2+k_{obs}^2\right)\hat{\eta}_1+\tau_h+\left(1+k_{obs}^2\right)\eta_1, \tag{5.386}$$

$\hat{\eta}_1$ is obtained by integrating $\dot{\hat{\eta}}_1$, and the differential Eq. (5.386) is constructed from Eq. (5.349). The set of control gains that render the best closed-loop performance are

$$\mathbf{K}_s = \{\ 3.0, \quad 0.007\ \} \text{ and } k_{obs} = 25.$$

Experiment 4: To establish robustness, the controller defined in Eq. (5.352) is tested with a higher payload of $m = 0.55$ [kg]. The same set of control gains in Experiment 3 are used to allow for a fair comparison.

Experiment 5: Experiment 3 is repeated again with no payload and the same set of control gains as in Experiment 3.

Subplots (a) and (b) in Figures 12 – 14 show the camera measured beam displacements at $x = 0.45L$ and $x = 0.85L$, respectively, measured in pixels to quantify and compare the vibration regulation performance of the controllers. Subplots (c) and (d) in Figures 12 – 14 show the hub angle and hub torque, respectively, for the different controllers implemented. Subplots (e) in Figures 12 – 14 show the time response of the first modal coordinate. Figure 13 (f) shows the behavior of the time derivative of the modal coordinate obtained by the backwards difference algorithm for the full-state feedback control defined in Eq. (5.384). Figure 14 (f) shows the time response of the first mode estimate defined in Eq. (5.386).

The plots indicate that the hub angular displacement regulation performances of all the controllers are comparable. However, both the full-state feedback and the observer-based modal controllers outperform the PD controller in vibration regulation. Specifically, from subplots (a) and (b) Figures 12 – 14, we observe that at $x = 0.45L$ and $0.85L$, the beam exhibits large displacements with a significant first flexible mode component with the PD controller; however, these displacements are very quickly reduced to less than ± 1 pixels within 4 [sec] and almost completely eliminated by 8 [sec]. The results of Experiments 4 and 5 are very close to Experiment 3, showing that the observer-based feedback controller is robust to large variations in payload mass.

Figure 12 shows that the first mode dominates the closed-loop system response for an angular displacement setpoint objective; hence, only the first flexible mode is incorporated into the modal controller design. Higher-order modes remain bounded, validating the theoretical prediction.

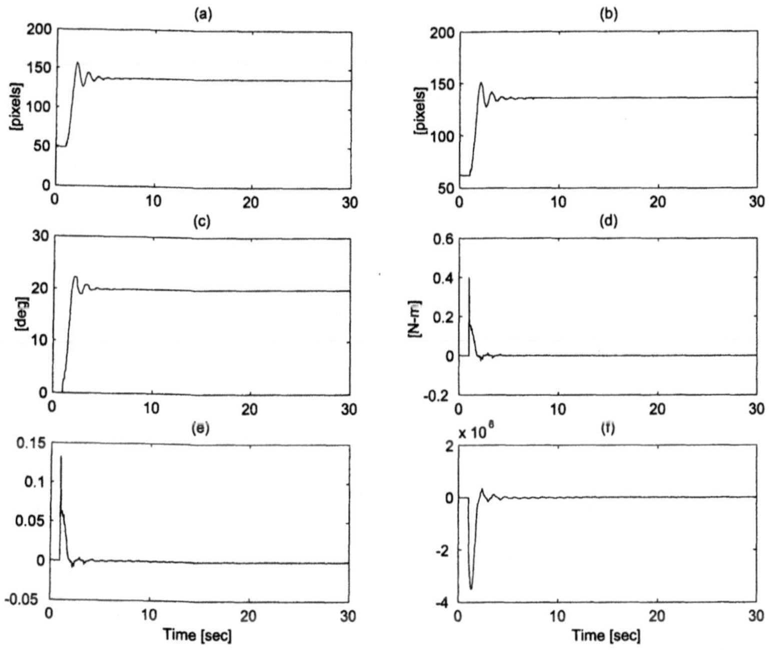

Figure 14 Experimental modal control with an observer for the modal coordinate time derivative: (a) Beam displacement at $x = 0.45L$, (b) Beam displacement at $x = 0.85L$, (c) Hub angle, (d) Applied hub torque, (e) Estimated modal coordinate $\hat{\eta}_1$, and (f) Estimated modal velocity $\dot{\hat{\eta}}_1$.

Chapter 6
ADAPTIVE CONTROL

Passive and EMK controllers have robustness and performance characteristics, respectively, that make them attractive solutions to a variety of problems. Adaptive control is designed to improve the robustness of EMK controllers without sacrificing performance. Adaptive control propagates parameter estimates via time integration, resulting in dynamic compensation. Thus, the theory and implementation of adaptive control is more complicated.

In this chapter, several previously developed EMK controllers are redesigned as adaptive controllers that compensate for parametric uncertainty. The EMK controllers must be linearly parameterizable to allow adaptive control development. While most of the EMK controllers provide exponential stability, the adaptive controllers only provide an asymptotic result. The time derivative of the Lyapunov functional for an adaptive controller is negative semi-definite. Thus, Barbalat's lemma must be invoked in the stability analysis.

6.1 Boundary Control

In this section, we design adaptive boundary controllers for the second order duct, cable, and nonlinear string and fourth order cantilevered beam models. Mechatronic systems are constructed to experimentally implement passive, EMK, and/or adaptive controllers for these systems. The performance and robustness of the controllers are compared experimentally.

6.1.1 Second Order Systems

Boundary Control of an Acoustic Duct

Revisiting the acoustic duct model of Eqs. (5.1) - (5.3), we design an adaptive boundary controller that provides asymptotic noise regulation and compensation for parametric uncertainty. The constant parameters of the duct and control speaker are estimated on-line.

Adaptive Control Law The controlled boundary dynamics for the duct model given in Eq. (5.5) can be linearly parameterized as follows

$$m\dot{\eta}(t) = -\mathbf{W}(u_{xt}(L,t), u_x(L,t), u_t(L,t), u(L,t))\,\mathbf{\Theta} + f(t), \tag{6.1}$$

where $\eta(t)$ is given in Eq. (5.4), the regression vector

$$\mathbf{W}(\cdot) = \left[\begin{array}{cccc} -\alpha u_{xt}(L,t) & u_x(L,t) & u_t(L,t) & u(L,t) \end{array} \right], \tag{6.2}$$

and the constant unknown parameter vector

$$\mathbf{\Theta} = \left[\begin{array}{cccc} m & \mathcal{B}A & B_d & K_s \end{array} \right]^T. \tag{6.3}$$

Based on the form of Eq. (6.1), the boundary control force is defined as

$$f(t) = \mathbf{W}(\cdot)\hat{\mathbf{\Theta}}(t) - \kappa\eta(t), \tag{6.4}$$

where κ is a positive constant feedback gain and

$$\hat{\mathbf{\Theta}}(t) = \left[\begin{array}{cccc} \hat{m}(t) & \widehat{\mathcal{B}A}(t) & \hat{B}_d(t) & \hat{K}_s(t) \end{array} \right]^T \tag{6.5}$$

is the dynamic, parameter estimate vector. The parameter estimate vector is updated according to the gradient update law

$$\dot{\hat{\mathbf{\Theta}}}(t) = -\mathbf{\Gamma}\mathbf{W}^T(\cdot)\eta(t), \tag{6.6}$$

where $\mathbf{\Gamma} \in \mathbb{R}^{4\times4}$ is a diagonal, positive-definite, adaptation gain matrix.

If we define $\tilde{\mathbf{\Theta}}(t) = \mathbf{\Theta} - \hat{\mathbf{\Theta}}(t)$ as the parameter estimation error vector, and substitute Eq. (6.4) into Eq. (6.1), we obtain the following closed-loop dynamics

$$m\dot{\eta}(t) = -\mathbf{W}(t)\tilde{\mathbf{\Theta}}(t) - \kappa\eta(t), \quad \dot{\tilde{\mathbf{\Theta}}}(t) = \mathbf{\Gamma}\mathbf{W}^T(\cdot)\eta(t), \tag{6.7}$$

where Eq. (6.6) has been used to obtain the parameter estimation error dynamics.

We assume existence of a solution for the dynamics given by Eqs. (5.1) - (5.3) under the proposed control, and that the distributed variables $u(x,t)$, and $u_t(x,t)$ belong to a space of sufficiently smooth functions.

Stability Analysis

Theorem 58 *The adaptive boundary control law given by (6.4) - (6.6) strongly asymptotically stabilizes the acoustic displacement as follows*

$$\lim_{t\to\infty} |u(x,t)| = 0 \quad \forall x \in [0,L]. \tag{6.8}$$

Proof: We augment the EMK Lyapunov functional in Eq. (5.8) as follows

$$V_a(t) = E_s(t) + E_c(t) + \frac{1}{2}m\eta^2(t) + \frac{1}{2}\tilde{\mathbf{\Theta}}^T(t)\mathbf{\Gamma}^{-1}\tilde{\mathbf{\Theta}}(t), \tag{6.9}$$

where the last term accounts for estimation error. This functional is bounded as follows

$$\lambda_1\left[E_s(t)+\eta^2(t)+\left\|\tilde{\Theta}(t)\right\|^2\right]\leq V_a(t)\leq\lambda_2\left[E_s(t)+\eta^2(t)+\left\|\tilde{\Theta}(t)\right\|^2\right], \quad (6.10)$$

where positive scalar constants

$$\begin{aligned}\lambda_1 &= \min\left\{1-\frac{4\beta_c\rho L}{\beta_s\min\{\rho,\mathcal{B}\}},\frac{m}{2},\frac{\lambda_{\min}\{\Gamma^{-1}\}}{2}\right\}>0,\\ \lambda_2 &= \max\left\{1+\frac{4\beta_c\rho L}{\beta_s\min\{\rho,\mathcal{B}\}},\frac{m}{2},\frac{\lambda_{\max}\{\Gamma^{-1}\}}{2}\right\}>1.\end{aligned} \quad (6.11)$$

After taking the time derivative of Eq. (6.9), and substituting Eqs. (5.1), (6.6), and (6.7) into the resulting expression, we obtain the following upper bound on the time derivative of $V_a(t)$

$$\dot{V}_a(t)\leq-\lambda_3\left(\eta^2(t)+E_s(t)\right)\triangleq-g_a(t), \quad (6.12)$$

where $g_a(t)$ is a non-negative scalar function, and

$$\lambda_3=\min\left\{\left(\kappa-\frac{\beta_s\mathcal{B}A}{2\alpha}\right),2\frac{\beta_c}{\beta_s}\right\}>0. \quad (6.13)$$

The time derivative

$$\dot{g}_a(t)=\lambda_3\left(2\eta^T(t)\dot{\eta}(t)+\dot{E}_s(t)\right). \quad (6.14)$$

From Ineqs. (5.11) and (5.19), we have $V_a(t)$ constant or decreasing $\forall t\in[0,\infty)$ and lower bounded by zero. It follows from Ineq. (5.11) that $E_s(t)$ and $\eta(t)$ are bounded $\forall t\in[0,\infty)$. Since $E_s(t)$ is bounded $\forall t\in[0,\infty)$, we know that the potential and kinetic energies defined in Eq. (5.9) are bounded. We now use Ineq. (3.5) to show that $u(x,t)$ is bounded $\forall t\in[0,\infty)$ and $\forall x\in[0,L]$. We assume $u_t(x,t)\in\mathcal{S}([0,L])$ so bounded kinetic energy means $u_t(x,t)\in\mathcal{L}_\infty$. We assume $u(x,t)\in\mathcal{S}_{A_0}([0,L])$ so $u_x(L,t)$ and $u_{xx}(x,t)$ $\forall x\in[0,L]$ are bounded $\forall t\in[0,\infty)$. Since $u_x(L,t)$ and $\eta(t)$ are bounded $\forall t\in[0,\infty)$, we use Eq. (5.4) to show that $u_t(L,t)$ is bounded $\forall t\in[0,\infty)$. Given $u_{xx}(x,t)\in\mathcal{L}_\infty([0,L])$, the field Eq. (5.1) implies that $u_{tt}(x,t)$ is bounded $\forall x\in[0,L]$ and $\forall t\in[0,\infty)$ and because $\dot{\eta}\in\mathcal{L}_\infty$ we obtain $u_{xt}(L,t)\in\mathcal{L}_\infty$. From Eq. (5.23) we know that $\dot{E}_s\in\mathcal{L}_\infty$. Thus all terms in $\dot{g}_a$ are bounded and from Barbalat's lemma,

$$\lim_{t\to\infty}E_s(t),\ |\eta(t)|=0. \quad (6.15)$$

Finally, we use Eq. (5.9) and Ineq. (6.10) to produce

$$\frac{\beta_s\mathcal{B}A}{2L}u^2(x,t)\leq\frac{\beta_s\mathcal{B}A}{2}\int_0^L u_s^2(s,t)\,ds\leq E_s(t)\leq\frac{1}{\lambda_1}V_a(t)\quad\forall x\in[0,L]. \quad (6.16)$$

Eq. (6.15) and the bound developed in Ineq. (6.16) yield the result of Eq. (6.8).
□

Experimental Setup Implementation of the EMK and adaptive control laws require the boundary measurements: *(i)* $u(L,t)$, the loudspeaker baffle displacement, *(ii)* $u_t(L,t)$, the baffle velocity, *(iii)* $u_x(L,t)$, the acoustic displacement gradient at $x = L$, and *(iv)* $u_{xt}(L,t)$, the time derivative of the displacement gradient at $x = L$. We use the acoustic pressure relationship $u_x(L,t) = P(t)/\mathcal{B}$ to measure $u_x(L,t)$, where $P(t)$ denotes the pressure measured by a microphone.

The experimental test-stand consists of a 0.381 [m] diameter, 3 [m] long PVC pipe with 0.41 [m] diameter and 300 [W] loudspeakers mounted at each end. An electret condenser microphone is placed very close to the loudspeaker baffle to measure the boundary pressure. A high-sensitivity accelerometer mounted on the polypropylene baffle of the loudspeaker measures loudspeaker baffle acceleration.

A Pentium 400 [MHz] PC running RT-Linux (a real-time operating system) hosts the control algorithm. A graphical user-interface provides an environment to write the control algorithm. The *MultiQ* I/O board provides for the data transfer between computer subsystem and the sensor-actuator subsystem. The current flowing through the actuator coil is sensed using a Hall-effect current sensor. The output of the current sensor is fed into an A/D channel of the I/O board. Two other A/D channels read in the accelerometer and microphone signals. Two D/A channels output voltages that drive the disturbance and actuator loudspeakers. These voltages go through two stages of amplification, consisting of standard OP07C operational amplifiers and Techron linear power amplifiers, for the first and second stages, respectively. The controller is implemented using a sampling period of 0.5 [m-sec]. The baffle velocity and displacement are computed by using integrating filters on the accelerometer signal, while the pressure signal derivative is obtained by using a second-order filtered backwards difference algorithm.

Experimental Results The objective of the experiment is to attenuate the acoustic noise in the duct. A sinusoidal disturbance voltage signal with a frequency sweep from 100 to 420 [Hz] in 16 [sec], and defined as

$$V_{dist} = 1.5\sin\left(2\pi\left(20t + 100\right)t\right) \ \text{[volts]},$$

is applied to the loudspeaker generating the acoustic disturbance.

First, the disturbance is applied in "open-loop" mode with the actuator loudspeaker completely de-energized. Figure 1 plots the pressure at the loudspeaker baffle measured by the microphone. It is observed that as the sinusoid approaches a duct resonant frequency, the sound magnitude increases rapidly, peaks at the resonant frequency, and then drops sharply as the disturbance frequency increases above the resonant frequency. Figures 2 and 3 show the open-loop frequency response of the boundary pressure and velocity, respectively. Six resonances are clearly visible.

Next, the adaptive controller defined in Eqs. (6.4) and (6.6) is implemented. The best performance of the controller is achieved with the control gains

$$\alpha = 2.0,\ \kappa = 1.0, \mathbf{\Gamma} = \text{diag}\left\{3.6\times 10^{-4}, 0.7, 12.2, 132.0\right\}.$$

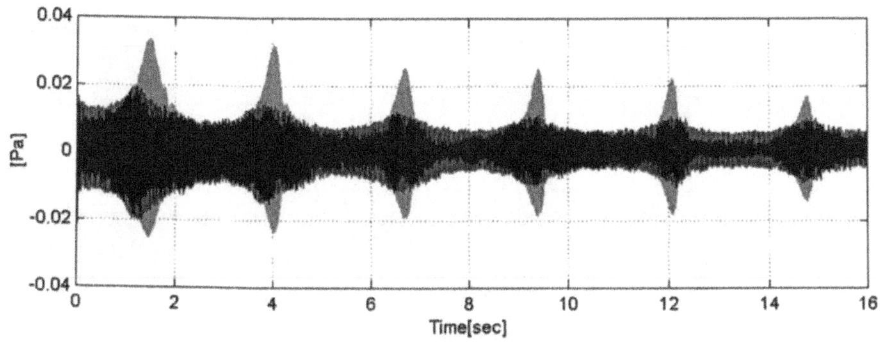

Figure 1 Experimental time response of the pressure at the loudspeaker baffle $\mathcal{B}u_x(L,t)$: Open-loop (gray line) and adaptive control (black line).

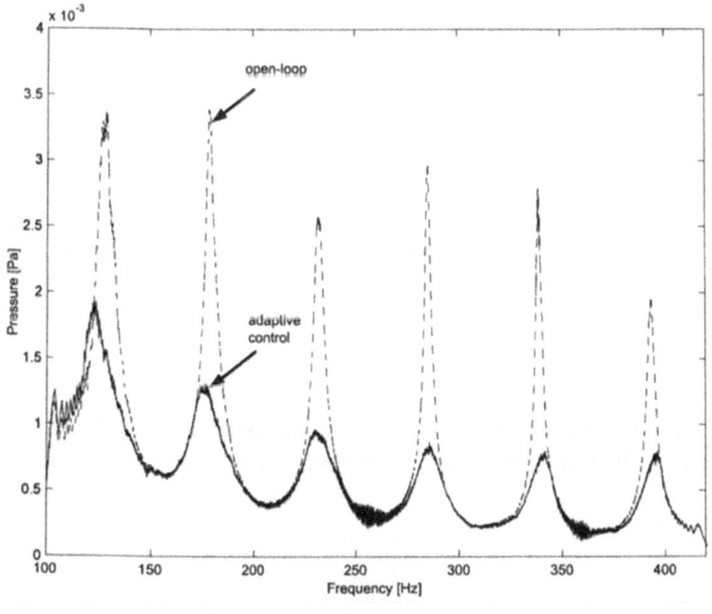

Figure 2 Experimental frequency response of the pressure at the loudspeaker baffle $\mathcal{B}u_x(L,t)$: Open-loop (dashed line) and adaptive control (solid line).

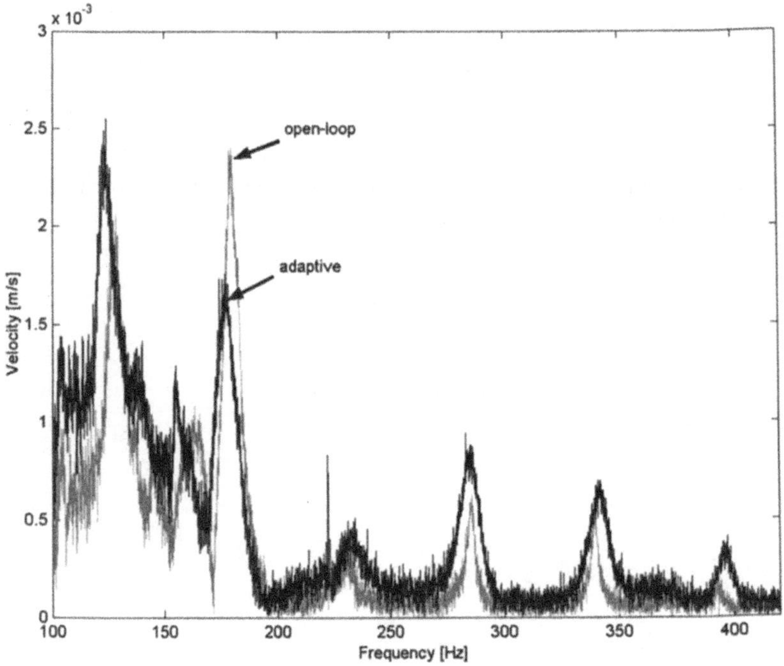

Figure 3 Experimental frequency response of the loudspeaker baffle velocity $u_t(L,t)$: Open-loop (gray line) and adaptive control (black line).

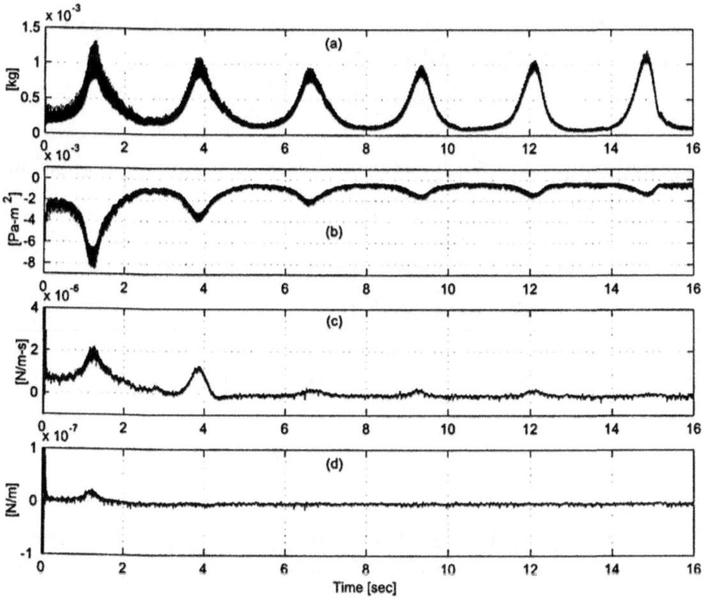

Figure 4 Adaptive control parameter estimates: (a) $\hat{m}(t)$, (b) $\widehat{BA}(t)$, (c) $\hat{B}_d(t)$, and (d) $\hat{K}_s(t)$.

The time response of the pressure at the loudspeaker baffle, and the parametric estimates are shown in Figures 1 and 4, respectively.

Figure 2 compares the open and closed-loop frequency response of the pressure at the loudspeaker baffle. The adaptive controller attenuates the peak response by over a factor of two when compared to the open-loop mode over the range of frequencies shown. Figure 3 illustrates the frequency response of the loudspeaker baffle velocity in the open-loop mode and with the adaptive controller. It is seen that at most resonant peaks, the loudspeaker baffle motion with the adaptive controller is larger than that in the open-loop mode. The controller causes relatively large loudspeaker baffle motion to drain acoustic energy out of the closed-loop system. This, in turn, allows the controller to achieve the desired acoustic noise regulation results. The parameter estimates remain bounded during closed loop operation as seen in Fig. 4.

6.1.2 Fourth Order Systems

Boundary Control of a Cantilevered Beam

Adaptive Control Derivation The EMK control law for the boundary controlled cantilevered beam given by Eq. (5.39) requires exact knowledge of some of the mechanical subsystem parameters. Now we redesign the controller to compensate for *constant* parametric uncertainty while asymptotically stabilizing the beam displacement. We define the adaptive control law

$$f = -\mathbf{W}\hat{\mathbf{\Theta}} - k_s\eta, \tag{6.17}$$

where $\eta(t)$ is defined in Eq. (5.40), $\mathbf{W}(t)$ and $\hat{\mathbf{\Theta}}(t)$ are a known regression matrix and a dynamic parameter estimate vector as follows:

$$\begin{aligned} \mathbf{W} &= \begin{bmatrix} -w_{xxxt}(L,t) & w_{xxx}(L,t) \end{bmatrix} & (6.18) \\ \hat{\mathbf{\Theta}} &= \begin{bmatrix} \hat{m} & \widehat{EI} \end{bmatrix}^T . & (6.19) \end{aligned}$$

We update the parameter estimate vector on-line according to

$$\dot{\hat{\mathbf{\Theta}}} = \mathbf{\Gamma}\mathbf{W}^T\eta \tag{6.20}$$

where $\mathbf{\Gamma}$ is a constant, diagonal gain matrix.

After differentiating Eq. (5.40) with respect to time, multiplying the resulting expression by m, and then substituting the right-hand side of Eq. (2.49) for $mw_{tt}(L,t)$, we have

$$m\dot{\eta} = \mathbf{W\Theta} + f \tag{6.21}$$

where $\mathbf{\Theta}$ is the unknown constant parameter vector defined as

$$\mathbf{\Theta} = \begin{bmatrix} m & EI \end{bmatrix}^T . \tag{6.22}$$

After substituting Eq. (6.17) into Eq. (6.21), we obtain the closed-loop boundary and estimator dynamics

$$\begin{aligned} m\dot{\eta} &= -k_s\eta + \mathbf{W}\tilde{\mathbf{\Theta}} & (6.23) \\ \dot{\tilde{\mathbf{\Theta}}} &= -\mathbf{\Gamma}\mathbf{W}^T\eta & (6.24) \end{aligned}$$

where $\tilde{\mathbf{\Theta}}(t) = \mathbf{\Theta} - \hat{\mathbf{\Theta}}(t)$ is the parameter estimate error vector and Eq. (6.20) has been used to obtain the parameter estimation error dynamics.

Theorem 59 *Given the field equation dynamics of Eq. (2.47) and the boundary conditions given by Eq. (2.48) and Eq. (2.49), the boundary controller given by Eq. (6.17) and Eq. (6.20) strongly asymptotically stabilizes the beam displacement as follows*

$$\lim_{t\to\infty} |w(x,t)| = 0 \qquad \forall x \in [0,L]. \tag{6.25}$$

Proof: We begin with the Lyapunov functional

$$V_a(t) = V(t) + \frac{1}{2}\tilde{\mathbf{\Theta}}^T(t)\mathbf{\Gamma}^{-1}\tilde{\mathbf{\Theta}}(t) \tag{6.26}$$

where $V(t)$ is defined in Eq. (5.44). If the design constant β of Eq. (5.46) is selected to be sufficiently small, we can formulate the following upper and lower bound on $V_a(t)$ defined in Eq. (6.26)

$$\lambda_{1a}\overline{V}_a(t) \leq V_a(t) \leq \lambda_{2a}\overline{V}_a(t) \tag{6.27}$$

where $\overline{V}_a(t)$ is defined as

$$\overline{V}_a(t) = \left[E_b(t) + \eta^2(t) + \left\|\tilde{\Theta}(t)\right\|^2\right], \tag{6.28}$$

for some positive constants λ_{1a} and λ_{2a} (Simple extension of Lemma 41). After differentiating Eq. (6.26) with respect to time and substituting Eq. (2.47), Eq. (2.48), and Eq. (6.23), we can obtain the following upper bound for the time derivative of $V_a(t)$

$$\dot{V}_a(t) \leq -\lambda_{3a}\left(E_b(t) + \eta^2(t)\right) = g(t) \tag{6.29}$$

for some positive constant λ_{3a} (Simple extension of Lemma 42). We can now use Eq. (6.27), Eq. (6.29), and standard smoothness arguments to state that all of the signals in the control of Eq. (6.17) and Eq. (6.20) are bounded $\forall t \in [0, \infty)$ and that all of the signals in the mechanical system given by Eq. (2.47) and Eq. (2.49) remain bounded $\forall t \in [0, \infty)$.

After differentiating the right-hand side of Eq. (6.29) with respect to time, we have

$$\dot{g}(t) = -\lambda_{3a}\left(\dot{E}_b(t) + 2\eta(t)\dot{\eta}(t)\right). \tag{6.30}$$

Since we illustrated how all of the system signals remain bounded $\forall t \in [0, \infty)$, we can use Eq. (5.55) and Eq. (6.23) to show that $\dot{E}_b(t)$ and $\dot{\eta}_1(t)$ are bounded $\forall t \in [0, \infty)$; hence, we can see from Eq. (6.30) that the time derivative of the right-hand of Eq. (6.29) is also bounded $\forall t \in [0, \infty)$. We now invoke Barbalat's Lemma to show that

$$\lim_{t\to\infty} E_b(t),\ \eta(t) = 0. \tag{6.31}$$

The inequality-type bound developed in Eq. (5.70) provides the strong asymptotic stability result given by Eq. (6.25).
□

Experimental Setup To experimentally test the proposed controllers, we construct a flexible beam hanging down from a rigid support structure. A metal cylinder is attached to a piezoelectric force sensor mounted on the bottom, free end of the beam. The axis of the beam is aligned with the center of gravity of the mass. Nonzero initial conditions are provided by mechanically latching the weight at an initial offset and releasing at $t = 0$. A pair of electromagnets apply a bi-directional force on the mass. The uncontrolled response from the latched initial displacement shown in Figure 5 has little damping.

In closed loop operation, high gain current feedback using Hall effect current sensors ensures that the actual magnetizing currents track the desired current trajectories. The signal from the force sensor's bridge circuit is scaled and filtered before being A/D converted. A laser sensor measures the boundary displacement. The controllers are implemented using a sampling period of 0.5 msec. Boundary velocity is obtained through backwards differencing of the displacement signal with the resulting signal being filtered by a second-order digital filter. The various parameters associated with the flexible beam model are calculated using standard test procedure and the engineering handbook tables (*i.e.*, $m = 0.56\ kg$, $L = 0.91\ m$, $EI = 1.27\ Nm^2$, and $\rho = 1.56\ kg/m$).

Experimental Results First, a passive damping control algorithm is implemented using the following control law

$$f = -k_d w_t(L,t) \tag{6.32}$$

where k_d is a scalar, positive constant control gain. The best regulating performance is achieved with $k_d = 100$. Figure 5 shows the much improved transient response under damper control.

Next, the EMK controller is implemented based on Eq. (5.39) but including integral action as follows

$$f = m\alpha u_{xxxt}(L,t) - EIu_{xxx}(L,t) - K_P\eta_1 - K_I \int_0^t \eta_1(t)\, dt. \tag{6.33}$$

where α, K_P, K_I are positive controller gains, and the variable $\eta_1(t)$ originally defined in (5.40) is now defined as follows

$$\eta_1 = u_t(L,t) - \alpha u_{xxx}(L,t) \tag{6.34}$$

where α is a positive controller gain [18]. The best performance is obtained with the following control gains

$$K_p = 100 \quad \alpha = 0.026 \quad K_I = 0.$$

Figure 5 shows more rapid transient decay than the damper control using similar control effort.

Finally, the adaptive controller is implemented without *a priori* knowledge of the system parameters Eq. (*i.e.*, the unknown parameter estimates are initialized to *zero*). The parameter estimates are computed on-line using trapezoidal integration. The best performance is achieved with the following control gains

$$\begin{array}{ll} K_p = 100 & \alpha = 0.02 \\ K_I = 0 & \mathbf{\Gamma} = \begin{bmatrix} 1 & 0 \\ 0 & 1 \end{bmatrix} \end{array}$$

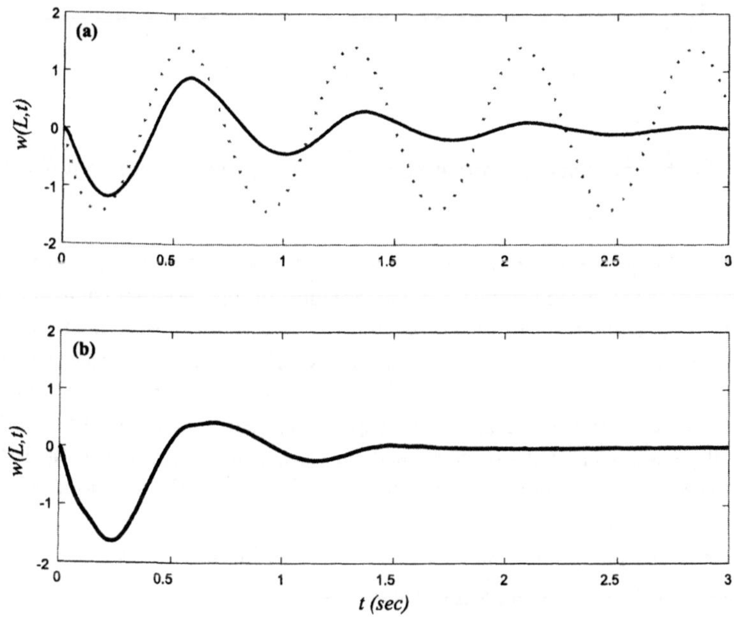

Figure 5 Cantilevered beam experimental endpoint displacement response: (a) Open loop (dotted) and damper controlled (solid), (b) EMK controlled

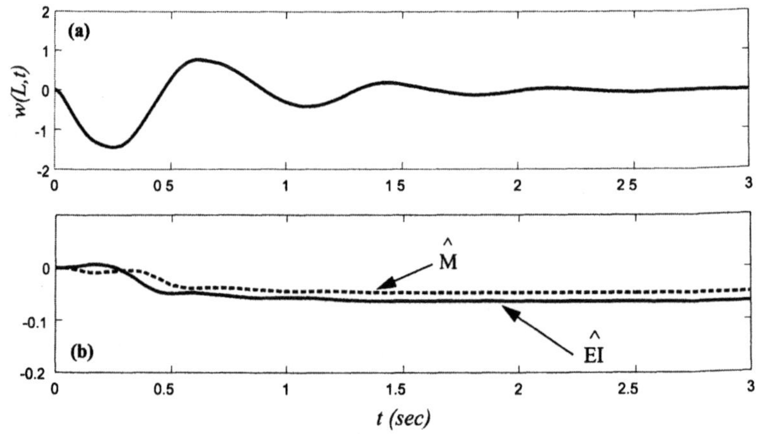

Figure 6 Cantilevered beam adaptive control: (a) Endpoint displacement, (b) Parameter estimates.

Figure 6 shows slightly reduced performance relative to EMK. As predicted by theory, the parameter estimates remain bounded but do not converge to their actual values.

The experimental results show that the maximum peak overshoot and the energy consumed for all three controllers are similar. The EMK controller exhibits the fastest settling time. The settling time of the adaptive controller is significantly better than that of damper controller despite the lack of *a priori* knowledge of the mechanical parameters. Persistent excitation would enable improved performance as the parameter estimates move toward their actual values and the response approaches that of the EMK controller.

6.1.3 Compensation for Geometric and Material Nonlinearities

Boundary Control of a Nonlinear String

Adaptive Control Derivation In this section, we redesign the EMK controller of Eq. (5.107) to compensate for parametric uncertainty. We rewrite the open-loop dynamics for $\eta(t)$ defined in Eq. (5.108) as follows

$$m\dot{\eta}(t) = \mathbf{W}(\cdot)\mathbf{\Theta} - P(t)\frac{w_x(L)}{\sqrt{1+w_x^2(L)}} + f(t) \tag{6.35}$$

where the matrix $\mathbf{W}(w_{xt}(L), w_t(L)) \in \Re^{1\times(p+1)}$ is given by

$$\mathbf{W}(\cdot) = \left[\frac{w_{xt}(L)}{\left(1+w_x^2(L)\right)^{3/2}}, \quad -Y(w_t(L))\right], \tag{6.36}$$

and the unknown parameter vector $\mathbf{\Theta} \in \Re^{p+1}$ is given by

$$\mathbf{\Theta} = \left[m, \ \phi^T\right]^T. \tag{6.37}$$

The adaptive controller consists of the control law

$$f(t) = -\mathbf{W}(\cdot)\hat{\mathbf{\Theta}} + P(t)\frac{w_x(L)}{\sqrt{1+w_x^2(L)}} - \left(k + \frac{k_r}{2}P(t)\right)\eta(t) \tag{6.38}$$

and the adaptation algorithm

$$\dot{\hat{\mathbf{\Theta}}}(t) = \mathbf{\Gamma}\mathbf{W}^T(\cdot)\eta(t) \tag{6.39}$$

that calculates the parameter estimate $\hat{\mathbf{\Theta}}(t)$ on-line. The adaptation gain matrix $\mathbf{\Gamma}$ is diagonal, positive-definite, and constant. After substituting Eq. (6.38) into Eq. (6.35), we obtain the closed-loop dynamics expression for $\eta(t)$ as follows

$$m\dot{\eta}(t) = -\left(k + \frac{k_r}{2}P(t)\right)\eta(t) + \mathbf{W}(\cdot)\tilde{\mathbf{\Theta}}(t) \tag{6.40}$$

where $\tilde{\mathbf{\Theta}}(t) = \mathbf{\Theta} - \hat{\mathbf{\Theta}}(t) \in \Re^{p+1}$ denotes the parameter estimation error vector.

To prove that all signals in the closed-loop system are bounded, we introduce the non-negative functional

$$V_{1a}(t) = V_1(t) + \frac{1}{2}\tilde{\mathbf{\Theta}}^T(t)\mathbf{\Gamma}^{-1}\tilde{\mathbf{\Theta}}(t) \tag{6.41}$$

where $V_1(t)$ is defined in Eq. (5.112). After differentiating $V_{1a}(t)$ with respect to time and substituting Eq. (6.40) and Eq. (6.39), we obtain $\dot{V}_{1a}(t) \leq 0$. Under the same assumptions used for the EMK control development, we can show that all the signals in the closed-loop system and the controller are bounded for all time.

To prove that the kinetic energy and the potential energy of the string-actuator system defined by Eq. (5.113) decay asymptotically to zero, we introduce a different functional

$$V_a(t) = V(t) + \frac{1}{2}\tilde{\mathbf{\Theta}}^T(t)\mathbf{\Gamma}^{-1}\tilde{\mathbf{\Theta}}(t) \tag{6.42}$$

where $V(t)$ is defined in Eq. (5.119). Following the same steps as with the EMK control development, we can show that $V_a(t)$ can be lower and upper bounded in the following manner

$$0 \leq \lambda_{1a}\left(E_a(t) + \eta^2(t) + \left\|\tilde{\mathbf{\Theta}}(t)\right\|^2\right) \leq V_a(t) \leq \lambda_{2a}\left(E(t) + \eta^2(t) + \left\|\tilde{\mathbf{\Theta}}(t)\right\|^2\right) \tag{6.43}$$

where λ_{1a}, λ_{2a} are positive constants that depend on μ_1, μ_2, $\lambda_{\min}(\mathbf{\Gamma}^{-1})$, and $\lambda_{\max}(\mathbf{\Gamma}^{-1})$. After differentiation and simplification of Eq. (6.42), we obtain

$$\dot{V}_a(t) \leq -g(t) \tag{6.44}$$

where $g(t)$ is defined in Eq. (5.141). Hence, we can repeat the same stability arguments used in proof of Theorem 40 to show that the kinetic energy and the potential energy of the string-actuator system decay asymptotically to zero.

Experimental Setup Figure 1 in Chapter 4 shows the schematic diagram of the experimental setup used to implement the controllers. The string tension, $P(t)$, required for implementation of the nonlinear controllers is obtained from the JR3 force sensor attached to the string at the pinned end. The time derivative of the string deflection angle is obtained using a backwards difference algorithm applied to the string deflection angle with the resulting signal filtered by a second-order digital filter. Trapezoidal integration implements the adaptive update laws Eq. (6.39).

Experimental Results Experiments are conducted to assess the performance of the proposed controllers. The transient response to a consistent initial displacement is studied. The control gains are tuned to provide the best response. The results can be compared with the open loop and damper control results shown in Fig. 2 of Chapter 4.

For small transverse displacement, the model-based controller of Eq. (6.38) becomes a *linear controller* as follows

$$f(t) = -mw_{xt}(L) + P_0 w_x(L) - \left(k + \frac{k_r}{2} P_0\right) \eta(t) \tag{6.45}$$

where P_0 is the mean tension and the actuator compensation term $Y(w_t(L))\phi$ is not used. The linear control is implemented with $P_0 = 29.75\ N$ and achieves the best regulation results with $k = 2.6$ and $k_r = 3.0$ shown in Figure 7. The inclusion of string slope feedback in the control causes more gantry motion and a corresponding increased (relative to open loop and damper control) damping, reducing the settling time to 6 sec.

Next we use the EMK control Eq. (5.107) with*

$$Y(\cdot) = [sgn(w_t(L)),\ w_t(L)] \qquad \phi = [F_s,\ F_d]^T \tag{6.46}$$

where F_s and F_d are the static and dynamic friction coefficients, respectively. The friction coefficients are experimentally determined to be $F_s = 1.05\ N$ and $F_d = 15.23\ N\sec/m$. The best regulation results are achieved with $k = 3.0$ and $k_r = 3.12$. Figure 8 shows the angular deflection $w_x(L)$, the gantry displacement $w(L)$, the time varying tension signal $P(t)$, and the control voltage resulting from the initial displacement. The nonlinear control response is only slightly improved compared to the linear control. The friction compensation, however, does prevent the gantry from sticking at a nonzero position as in Fig. 7.

*The sign function $sgn(x)$ is defined as follows

$$sgn(x) = \begin{cases} 1 & \text{if } x > 0 \\ 0 & \text{if } x = 0 \\ -1 & \text{if } x < 0. \end{cases}$$

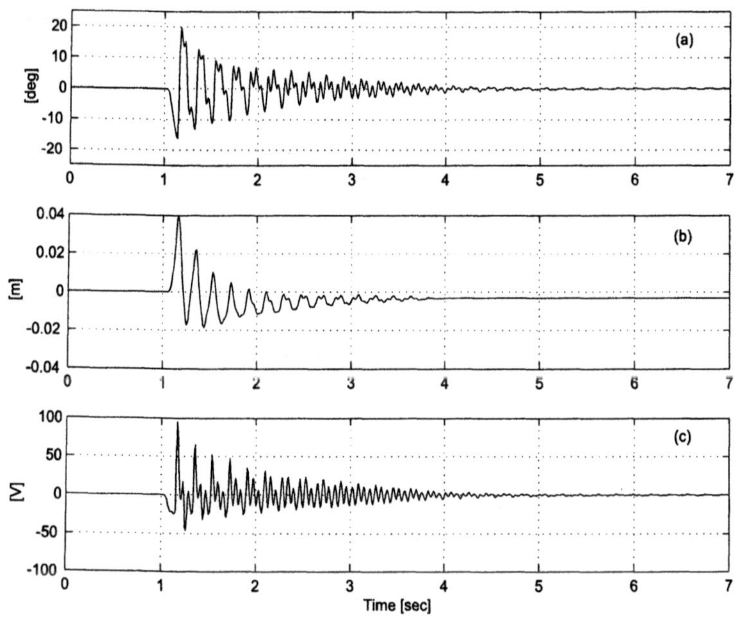

Figure 7 Experimental results for the linear string boundary controller: (a) string deflection angle $w_x(L,t)$, (b) gantry displacement $w(L,t)$, and (c) control voltage

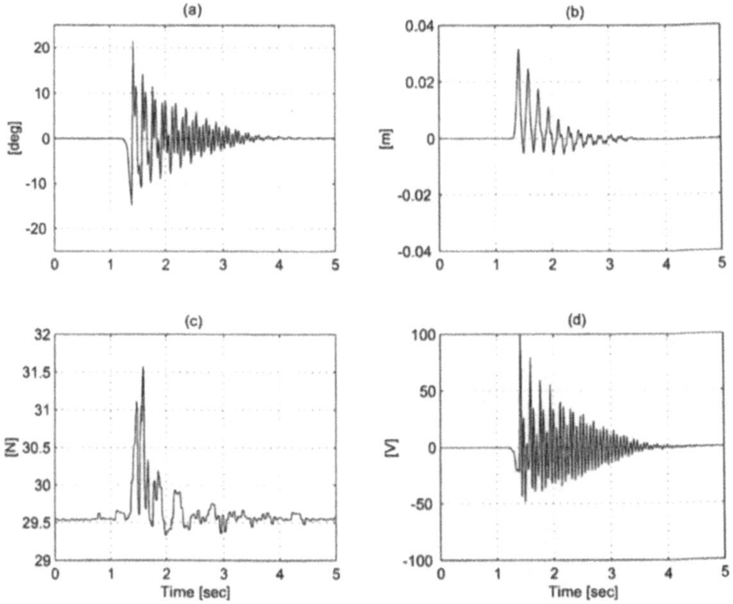

Figure 8 Experimental results for the nonlinear string EMK controller: (a) string deflection angle $w_x(L,t)$, (b) gantry displacement $w(L,t)$, (c) tension $P(t)$, and (d) control voltage.

Finally, the adaptive controller defined by Eq. (6.38) and the update laws given by Eq. (6.39) are implemented with the performance shown in Fig. 9. The parameter estimates are initialized to 25% of their nominal values. The best regulation results are achieved with $k = 3.0$, $k_r = 2.97$, and $\mathbf{\Gamma} = diag\{2.55, 50, 0.69\}$. Again, the response is similar to the previous two controllers. The estimated parameters converge to final values but not their exact values. Persistent excitation via sinusoidal or repeated impulse inputs would move the estimates to their actual values.

6.1.4 Effects of Nonuniform Parameter Distributions

Boundary Control of a Cable

Adaptive Control Development In this subsection, we redesign the exact model knowledge boundary controller for a cable given in Eq. (5.153) to compensate for *constant* parametric uncertainty while asymptotically stabilizing the mechanical model given by Eq. (2.168) through Eq. (2.170). The adaptive control law is

$$f = -\mathbf{W}\hat{\mathbf{\Theta}} - k_s\eta, \tag{6.47}$$

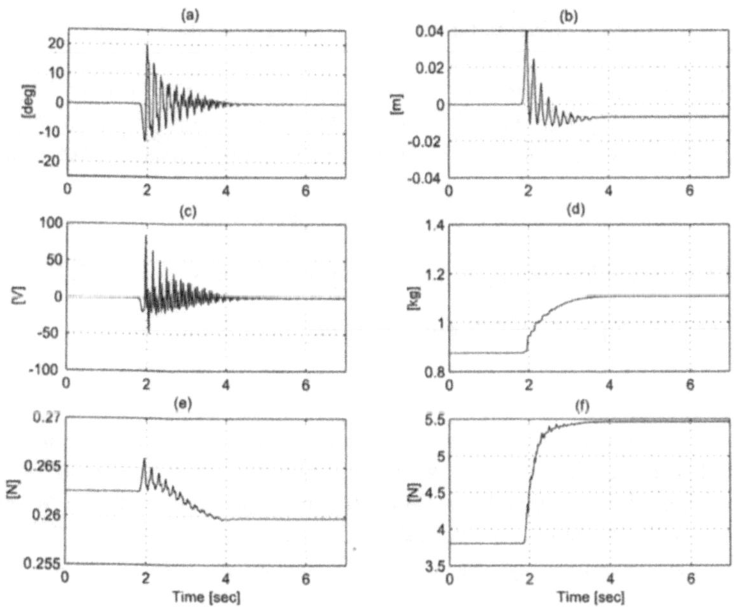

Figure 9 Experimental results for the nonlinear string adaptive controller: (a) string deflection angle $w_x(L,t)$, (b) gantry displacement $w(L,t)$, (c) control voltage, (d) mass estimate $\hat{m}(t)$, (e) static friction coefficient estimate $\hat{F}_s(t)$, and (f) dynamic friction coefficient estimate $\hat{F}_d(t)$.

where $\mathbf{W}(t)$ is a known regression matrix and $\hat{\mathbf{\Theta}}(t)$ is a dynamic parameter estimate vector:

$$\mathbf{W} = \left[\begin{array}{ccc} w_{st}(1,t) & -w_s(1,t) & -Y(w_t(1,t)) \end{array}\right], \qquad \hat{\mathbf{\Theta}} = \left[\begin{array}{ccc} \hat{m} & \widehat{p(1)} & \hat{\phi}^T \end{array}\right]^T. \quad (6.48)$$

The parameter estimate vector introduced in Eq. (6.47) is updated on-line according to

$$\dot{\hat{\mathbf{\Theta}}} = \mathbf{\Gamma}\mathbf{W}^T\eta, \quad (6.49)$$

where $\mathbf{\Gamma}$ is a constant, diagonal gain matrix with elements $\mathbf{\Gamma}_i > 0$.

Theorem 60 *Given the field equation of Eq. (2.168) and the boundary conditions given by Eq. (2.169) and Eq. (2.170), the boundary controller given by Eqs. (6.47) and (6.49) ensures that the cable displacement is strongly asymptotically stabilized as follows:*

$$\lim_{t\to\infty} |w(s,t)| = 0 \qquad \forall s \in [0,1], \quad (6.50)$$

provided the controller gain k_s defined in Eq. (6.47) satisfies Eq. (5.155).

Proof: The proof follows that of Theorem 58.
□

Experimental Results The proposed controller is implemented on the same experimental setup as the nonlinear string experiment but with a long, heavy braided polyester rope with significant sag (low tension). The rope is pinned at one end and connected to the horizontally-translating gantry at the other end. The parameter values for the mechanical system are determined via standard test procedures to be as follows:

Dimensional: $M = 3.229\ kg, \quad L = 2.69\ m, \quad \rho A = 0.085\ kg/m, \quad P_0 = 0.127\ N$

Nondimensional: $m = 14.1, \quad p_0 = 0.056, \quad p(1) = 0.503.$

Two experiments are conducted to test the performance of the proposed controllers. First, the EMK controller given by Eq. (5.153) is implemented using the following gains:

$$k_s = 5.0 \quad \alpha = 4.0 \quad Y(w_t(1,t))\phi = 0,$$

where we have added the additional control gain α in the definition of η as in Eq. (5.4). The experimental results in Fig. 10 show that the cable system exhibits excellent transient response under the proposed controller. The uncontrolled swinging of the cable to the same input is shown for comparison purposes. The gantry position and the voltage signals are also shown. It should also be noted that the $Y(w_t(1,t))\phi$ term could have been used to model static friction on the rail of the actuator; however,

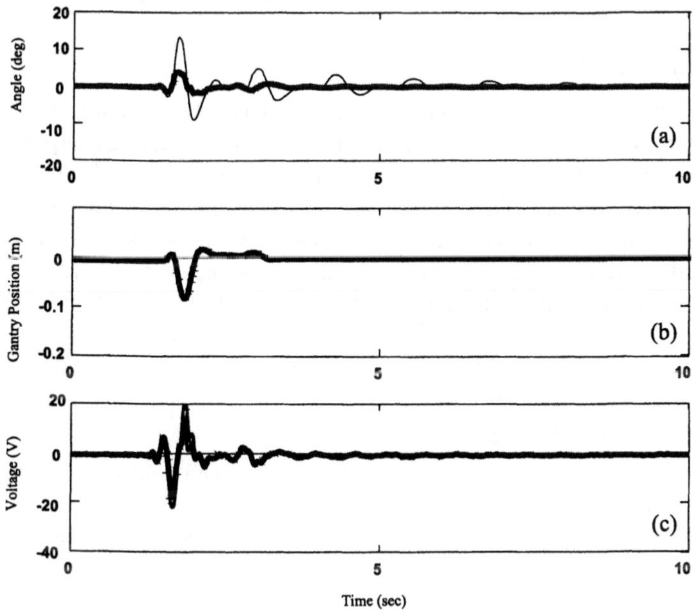

Figure 10 Experimental results for EMK boundary cable controller: (a) Cable departure angle $w_x(L,t)$; (b) Gantry displacement $w(x,t)$; (c) Control voltage.

the feedback portion of the controller seemed to adequately compensate for friction in this case.

The adaptive version of Eq. (5.153) is implemented with the update law given by Eq. (6.49) and the parameter estimates initialized to 80% of their nominal values. The best performance is achieved using the following gains:

$$k_s = 5.0 \quad \alpha = 4.0 \quad Y(w_t(1,t))\phi = 0 \quad \mathbf{\Gamma} = diag\{5, 2\}.$$

Figure 11, shows the transient performance of the adaptive controller to the disturbance and the uncontrolled swinging of the cable for the same input. The parameter estimates $\hat{m}$ and $\hat{p}(1)$ remain bounded during closed-loop operation.

6.2 Domain Control

Now we return to the domain control problem of isolating a controlled subdomain from bounded disturbances in an adjacent subdomain. We saw in Chapter 5 that with exact parameter knowledge we can transfer noise and vibration from the controlled subdomain to the uncontrolled subdomain and prevent disturbances in the uncontrolled subdomain from propagating to the controlled subdomain. In this section, we show that adaptive controllers can provide strong asymptotic regulation for the controlled subdomain and estimate the system parameters. Thus, the system parameters do not have to be known à priori. Retuning of the controller due to slow parameter changes is not required. Additionally, the time-varying parameter estimates can be used to monitor the system health.

6.2.1 Acoustic Noise Isolator

The adaptive noise isolator asymptotically stabilizes the controlled subdomain displacement and estimates the unknown parameter vector $\mathbf{\Theta}$. The controlled subdomain and adjacent uncontrolled subdomain are governed by Eqs. (2.35) - (2.39). We define the parameter estimate vector

$$\hat{\mathbf{\Theta}}(t) = \left[\hat{c}, \hat{k}, \hat{b}, \hat{A}, \hat{m}\right]^T, \tag{6.51}$$

where $\hat{m}$, $\hat{c}$, and $\hat{k}$ are the speaker mass over $\mathcal{B}$, damping, and stiffness estimates, respectively, $\hat{A}$ is the duct area estimate, and $\hat{b}$ is the estimate of $\frac{A_T}{\mathcal{B}}$. Next, we prove that an adaptive controller can asymptotically stabilize the quiet subdomain and maintain stability of the noisy subdomain using bounded control force.

Theorem 61 *The bounded control law*

$$f = \mathbf{W}\hat{\mathbf{\Theta}} - k_p\eta \tag{6.52}$$

and adaptation law

$$\dot{\hat{\mathbf{\Theta}}} = -\mathbf{\Gamma}\mathbf{W}^T\eta, \tag{6.53}$$

where $\mathbf{\Gamma} = \mathbf{\Gamma}^T > 0$, *ensure strong asymptotic stability for* $x \in [0, a]$.

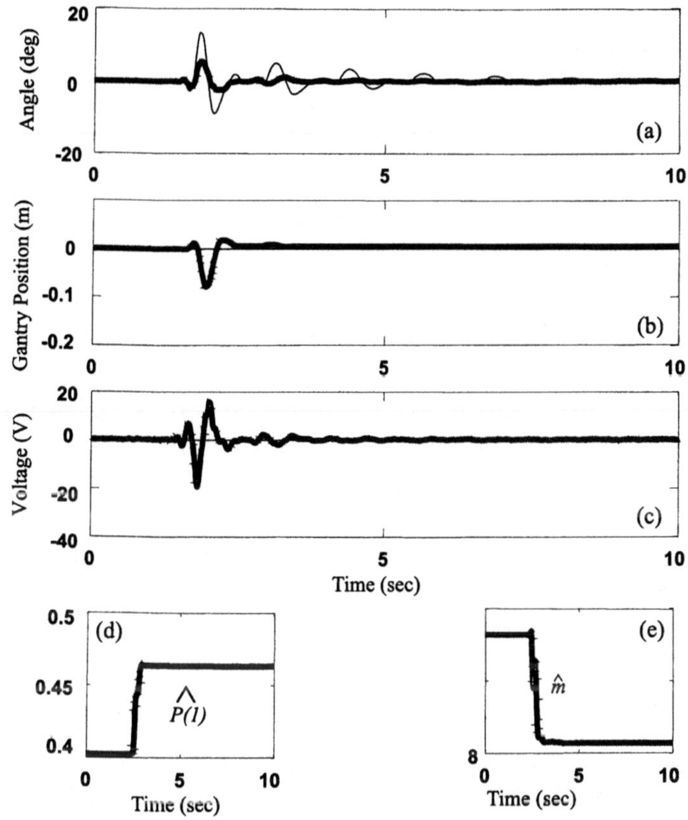

Figure 11 Experimental results of adaptive boundary cable control: (a) Cable departure angle $w_x(1,t)$; (b) Gantry displacement $w(1,t)$; (c) Control voltage; (d) Tension estimate $\hat{p}(1)$; (e) Mass estimate $\hat{m}$.

Proof: Substitution of Eq. (6.52) into Eq. (5.266) produces

$$m\dot{\eta}(t) = -k_p\eta + \mathbf{W}\tilde{\mathbf{\Theta}}, \tag{6.54}$$

where the parameter estimation error

$$\tilde{\mathbf{\Theta}}(t) = \hat{\mathbf{\Theta}} - \mathbf{\Theta}.$$

We define the Lyapunov functional

$$V_{ca}(t) = E_b^- + E_c^- + \frac{1}{2}m\eta^2 + E_a > 0, \tag{6.55}$$

where

$$E_a(t) = \frac{1}{2}\tilde{\mathbf{\Theta}}^T\mathbf{\Gamma}^{-1}\tilde{\mathbf{\Theta}},$$

and, as in the EMK proof,

$$0 < \lambda_{1a}\left(E_b^- + \eta^2 + \left\|\tilde{\mathbf{\Theta}}\right\|^2\right) \le V_{ca}(t) \le \lambda_{2a}\left(E_b^- + \eta^2 + \left\|\tilde{\mathbf{\Theta}}\right\|^2\right), \tag{6.56}$$

where λ_{1a} and λ_{2a} are positive constants. Repetition of the simplifying steps in the EMK proof and substitution of the adaptive law (6.53) yields

$$\dot{V}_{ca} \le -\lambda_3\left(E_b^- + \eta^2\right) = -g(t) \le 0 \tag{6.57}$$

Eqs. (6.55) and (6.57) mean that $V_{ca}(t) \le V_{ca}(0)$, or $V_{ca}(t) \in \mathcal{L}_\infty$. Therefore, η, $\tilde{\mathbf{\Theta}}$, $\hat{\mathbf{\Theta}}$, E_b^-, and $\int_0^a u_t^2 dx \in \mathcal{L}_\infty$. Assuming $u_t(x,t) \in \mathcal{S}([0,a])$ and $u_t(x,t) \in \mathcal{S}_{A_0}([0,a])$ implies that $u_t(x,t) \in \mathcal{L}_\infty([0,a])$ and $p_t^- \in \mathcal{L}_\infty$. From

$$\frac{u^2(x,t)}{a} \le \int_0^a u_x^2 dx \in \mathcal{L}_\infty([0,a]),$$

and assuming $u(x,t) \in \mathcal{S}_{A_0}([0,a])$ we know that $p^-(t) \in \mathcal{L}_\infty$ and $p_x(x,t) \in \mathcal{L}_\infty([0,a])$, which implies that $u_{tt}(x,t) \in \mathcal{L}_\infty([0,a])$. From Lemma 53, $u(x,t) \in \mathcal{L}_\infty([a,L])$ and, assuming $u_t(x,t) \in \mathcal{S}([a,L])$, $u_t(x,t) \in \mathcal{L}_\infty([a,L])$. If we assume $u(x,t), u_t(x,t) \in \mathcal{S}_{A_0}([0,a])$ then $u_t(x,t)$, $p(x,t)$, and $p_t(x,t) \in \mathcal{L}_\infty([a,L])$. Thus from Eqs. (5.267) and (6.52), $\mathbf{W}(t)$, $f(t)$, and $\dot{\eta}(t) \in \mathcal{L}_\infty$ and all signals are bounded.

Using Barbalat's lemma and

$$\dot{g} = -\lambda_3\left(\dot{E}_1 + \eta\dot{\eta}\right) \in \mathcal{L}_\infty,$$

we obtain $g \to 0$ as $t \to \infty$, which means that $u(x,t) \to 0$ asymptotically fast.
□

Figure 12 shows the duct experiment used to implement and test the isolation controllers. The test stand has disturbance speakers on both ends of the duct and

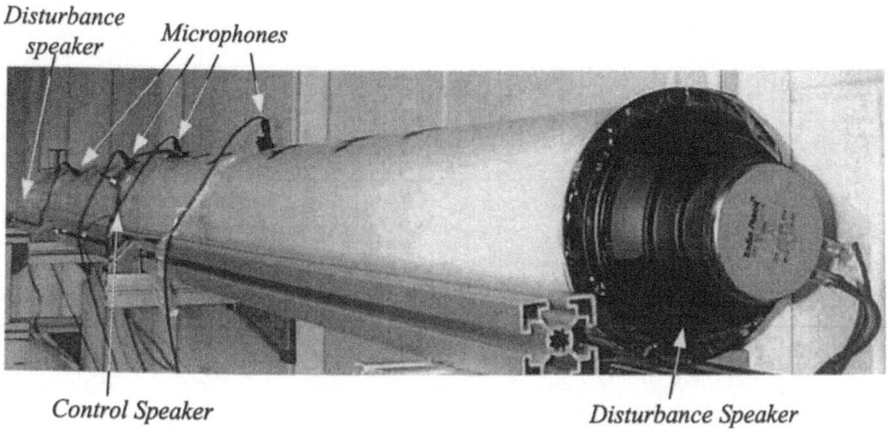

Figure 12 Noise isolation experimental setup

sensing microphones and a control speaker in the middle. In the experiments, one disturbance speaker provides a chirp disturbance (200Hz to 500Hz in 10 seconds with an amplitude of 1 V) to either the quiet or noisy side. Microphones measure the pressure on both sides of the actuating speaker, corresponding to the quiet $p^-(t)$ and noisy $p^+(t)$. A small accelerometer attaches to the actuating speaker cone to measure $u_{tt}(a,t)$. A combination of filters integrate the accelerometer output to provide velocity $u_t(a,t)$ and displacement $u(a,t)$ of the speaker. The actuating speaker is mounted inside the duct facing the controlled subdomain and is driven from computer through a MultiQ board. The control algorithm is implemented using a Simulink^{TM} block diagram on a mechatronics workstation running the Linux operating system with Real-Time Linux Target (RTLT) software.

Figure 13 (a) and (c) show the noisy pressure $p^+(t)$ FFT and time response, respectively, to a noisy-side chirp disturbance. In agreement with theory, the plots show that the closed loop noisy side response is bounded but not better than open loop. The quiet side response to the noisy side disturbance shown in Figs. 13 (b) (FFT) and (d), however, reduces by 4.4 to 9.2 dB at peak resonances under closed loop control. Some noise passes around and through the speaker so the exponential decay of quiet side noise predicted by theory does not occur. Figure 13 (e) shows the speaker velocity integrated from the accelerometer. The open and closed loop responses have similar amplitudes. The system model indicates that if the speaker displacement is regulated to zero, no noise will pass from the noisy side to the quiet side. In the experiment, however, noise propagates around and through the duct structure. The EMK controller damps this noise, causing significant speaker motion. The control voltage in Fig. 13 (f) appears smooth with reasonable magnitude.

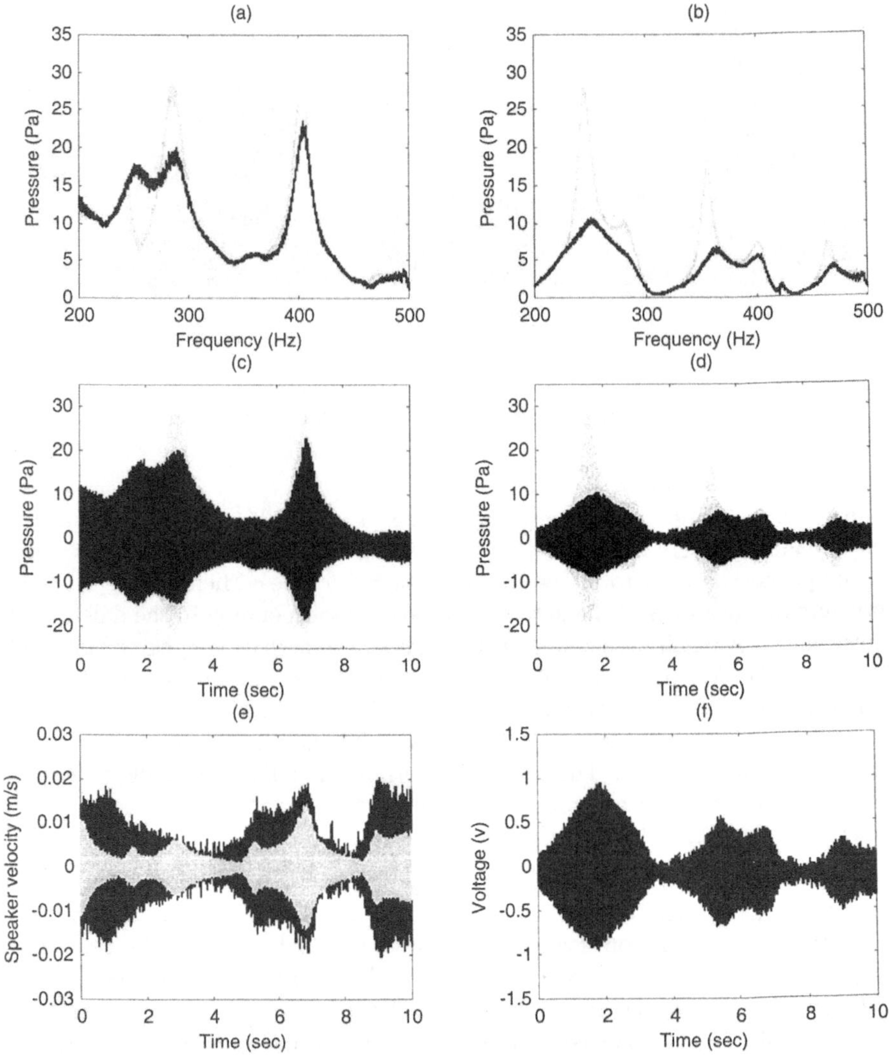

Figure 13 Experimental EMK response for noisy duct disturbance (open-loop: gray; closed-loop: black). (a) FFT of noisy duct microphone $p^+(t)$; (b) FFT of quiet duct microphone $p^-(t)$; (c) $p^+(t)$; (d) $p^-(t)$; (e) Speaker velocity $u_t(a,t)$; (f) Control voltage.

The signals plotted in Fig. 14 are the same as Fig. 13 except that the acoustic disturbance comes from the quiet side. The theoretical development assumes no disturbances on the quiet side and therefore does not predict the system response. Fig. 14 (a) - 14 (d) show however, that the EMK controller significantly reduces the resonant peaks on the quiet side by dissipating the acoustic energy and pumping it to the noisy side. The speaker moves considerably more than in open loop to facilitate this energy transfer. Again, the quiet side peaks are attenuated by 5.8 to 9.1 dB.

Figure 15 shows the adaptive controller response to the noisy side disturbance. Figure 16 shows the parameter estimates converging from zero initial conditions to final conditions. Initially, the feedforward terms are small because they multiply the small parameter estimates. The performance improves markedly as the parameters converge and eventually surpasses the EMK performance for the last resonant peak near 460 Hz. The parameter estimates are well behaved and bounded. The normalized RMS of the parameters is also shown in Fig. 16 (f).

Figure 17 shows similar trends as Fig. 15 for the adaptive control response to a quiet side disturbance. The parameters start at zero and remain bounded. The closed loop performance as measured by the quiet side peak responses improves as the parameters converge.

6.2.2 Axially Moving Beam Isolator

Adaptive control law

Recall the vibration isolating controller for the axially moving beam model given in Eqs. (2.58) - (2.63) . The control objective is to drive the beam displacement $w(x,t)$ for $x \in (0,a)$ to zero with a bounded control torque τ_c. In Chapter 5, we design an exponentially regulating control law for the system based on exact model knowledge. The EMK control law Eq. (5.310) depends on exact knowledge of the system parameters EI and J. To compensate for unknown or slowly time-varying parameters, we develop an adaptive controller. The open loop dynamics for the variable $\eta(t)$ are rewritten as

$$J\dot{\eta} = \tau_c + \mathbf{W}\boldsymbol{\Theta}, \tag{6.58}$$

where

$$\boldsymbol{\Theta} = \left[\; EI, \;\; J \;\right]^T, \; \mathbf{W} = \left[\; w_{xx}^{+} - w_{xx}^{-}, \;\; w_{xxt}^{-} \;\right].$$

The adaptive controller asymptotically stabilizes the controlled subdomain displacement using a bounded control torque and estimates the unknown parameter vector $\boldsymbol{\Theta}$.

Theorem 62 *The bounded adaptive control law*

$$\tau_{ca} = -\mathbf{W}\hat{\boldsymbol{\Theta}} - k_a \eta, \tag{6.59}$$

where

$$\hat{\boldsymbol{\Theta}} = \left[\; \hat{EI} \;\; \hat{J} \;\right]^T,$$

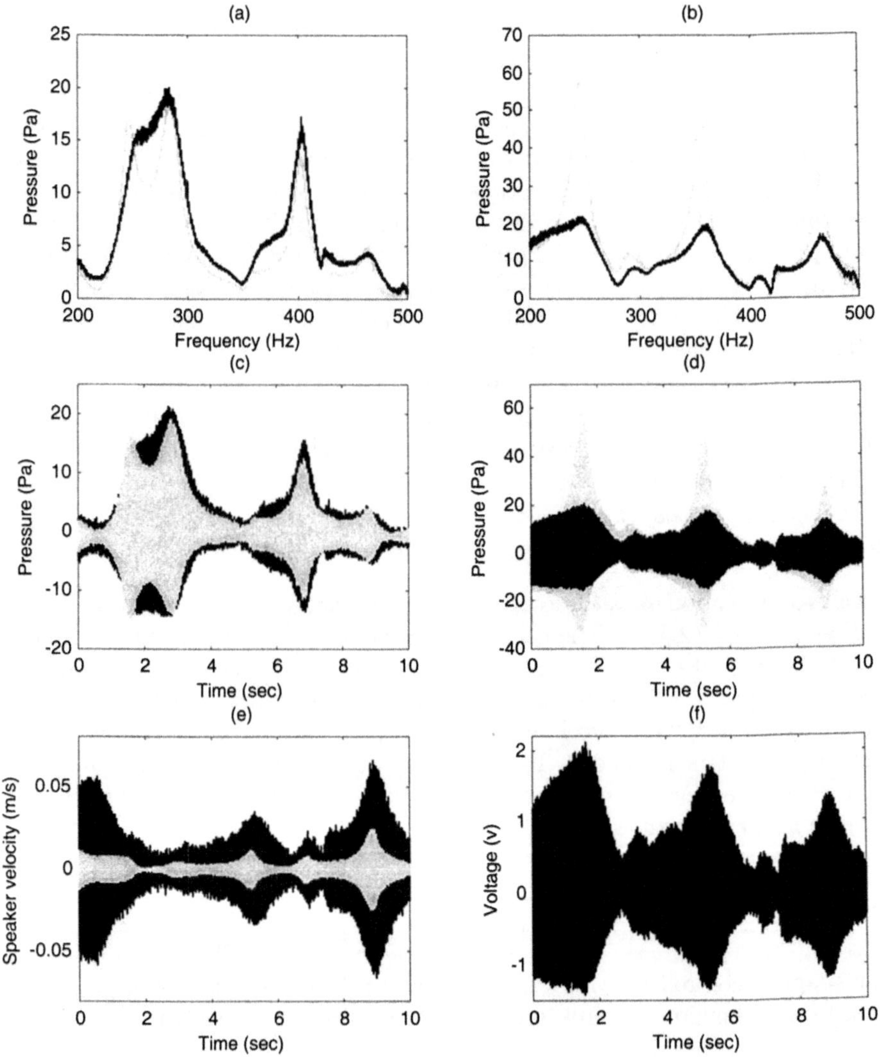

Figure 14 Experimental EMK response for quiet duct disturbance (open-loop: gray; closed-loop: black). (a) FFT of noisy duct microphone $p^+(t)$; (b) FFT of quiet duct microphone $p^-(t)$; (c) $p^+(t)$; (d) $p^-(t)$; (e) Speaker velocity $u_t(a,t)$; (f) Control voltage.

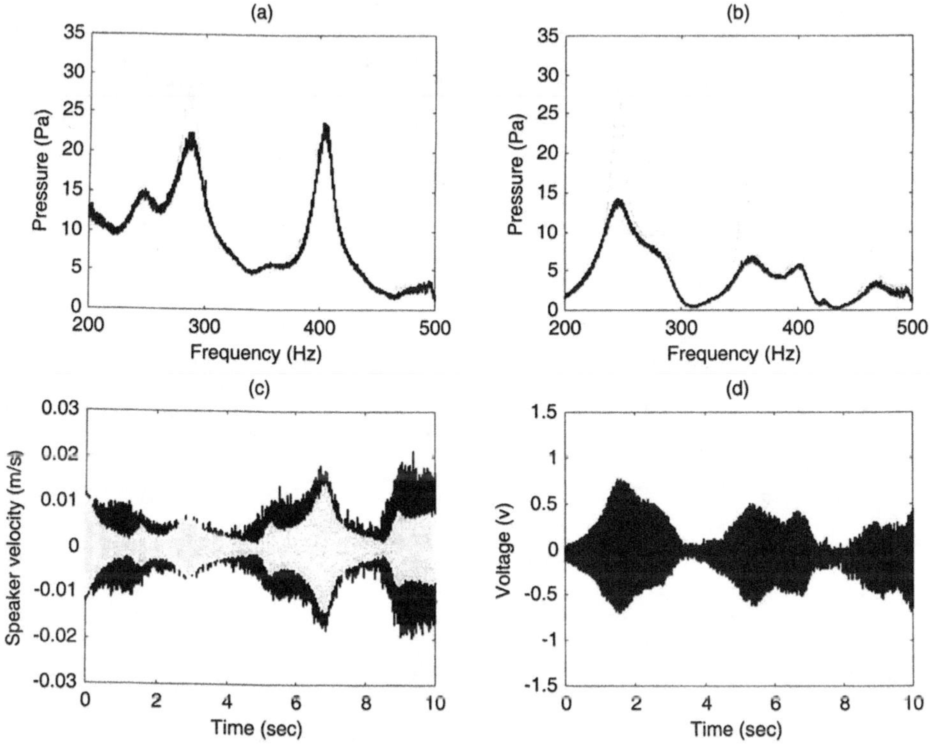

Figure 15 Experimental adaptive response for noisy duct disturbance (open-loop: gray; closed-loop: black). (a) FFT of noisy duct microphone $p\,(a^{+},t)$; (b) FFT of quiet duct microphone $p\,(a^{-},t)$; (c) Speaker velocity $u_t\,(a,t)$; (d) Control voltage.

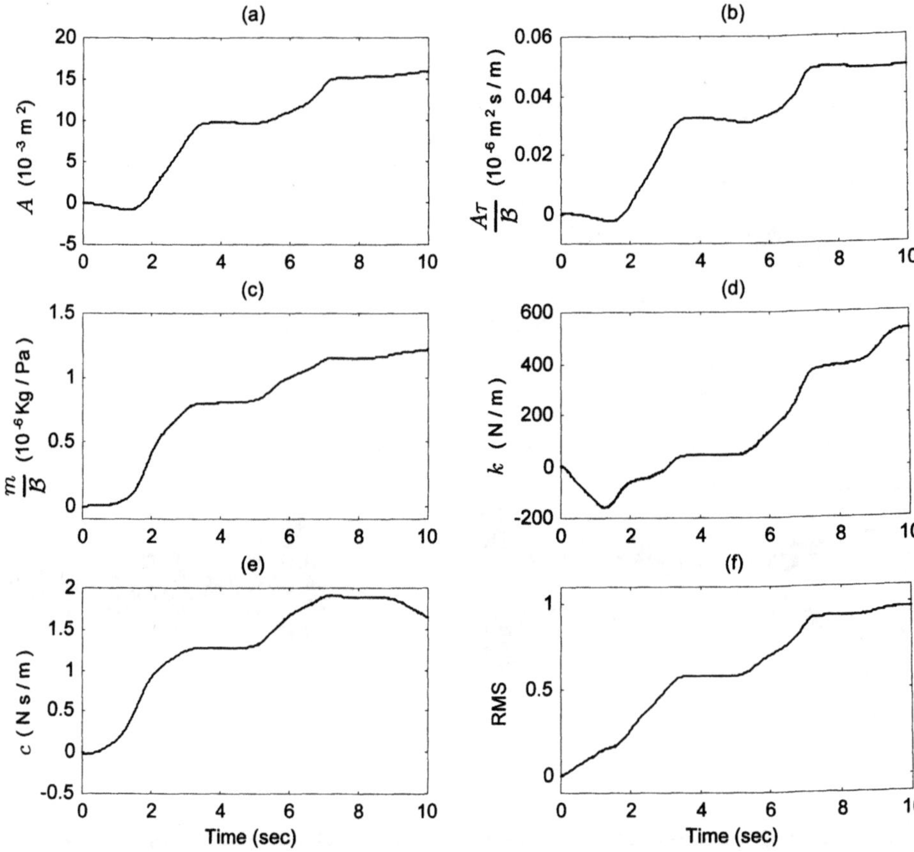

Figure 16 Parameter adaptation for noisy duct disturbance: (a) A; (b) $\frac{A\tau}{B}$; (c) $\frac{m}{B}$; (d) k; (e) c; (f) Normalized parameter RMS.

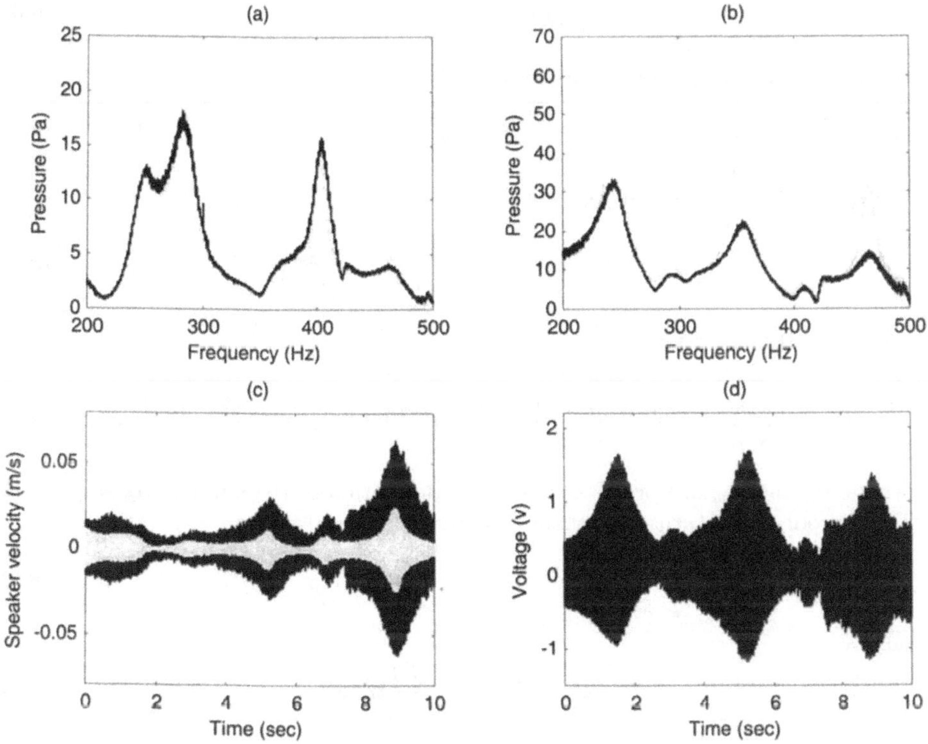

Figure 17 Experimental adaptive response for quiet duct disturbance (open-loop: gray; closed-loop: black). (a) FFT of noisy duct microphone $p^{+}(t)$; (b) FFT of quiet duct microphone $p^{-}(t)$; (c) Speaker velocity $u_t(a,t)$; (d) Control voltage.

with the adaptive update law:

$$\dot{\hat{\mathbf{\Theta}}} = \mathbf{\Gamma}\mathbf{W}^T\eta, \tag{6.60}$$

where $\mathbf{\Gamma} = \mathbf{\Gamma}^T > 0$, *ensure* $w(x,t) \to 0$, *for* $x \in [0,a]$ *as* $t \to \infty$.

Proof: Substitution of Eq. (6.59) into Eq. (6.58) produces

$$J\dot{\eta}(t) = -k_a\eta - \mathbf{W}\tilde{\mathbf{\Theta}}, \tag{6.61}$$

where the parameter estimation error

$$\tilde{\mathbf{\Theta}} = \hat{\mathbf{\Theta}} - \mathbf{\Theta}. \tag{6.62}$$

We define the Lyapunov functional

$$V_{ca}(t) = E_b^-(t) + E_c^-(t) + \frac{1}{2}J\eta^2 + \frac{1}{2}\tilde{\mathbf{\Theta}}^T\mathbf{\Gamma}^{-1}\tilde{\mathbf{\Theta}} > 0, \tag{6.63}$$

and, as in the EMK proof,

$$0 \leq \lambda_{1a}\left(E_s^- + \|\eta\|^2 + \left\|\tilde{\mathbf{\Theta}}\right\|^2\right) \leq V_{ca}(t) \leq \lambda_{2a}\left(E_s^- + \|\eta\|^2 + \left\|\tilde{\mathbf{\Theta}}\right\|^2\right), \tag{6.64}$$

where λ_{1a} and λ_{2a} are positive constants. Repetition of the simplifying steps in the EMK proof and substitution of the adaptive law (6.60) yields

$$\dot{V}_{ca}(t) \leq -\lambda_{3a}\left(E_b^- + \eta^2\right) = -g \leq 0, \tag{6.65}$$

where

$$\lambda_{3a} = \min\left(k_a - \frac{EI}{2}\alpha,\ 2\beta\right). \tag{6.66}$$

Eqs. (6.63) and (6.65) mean that $V_{ca}(t) < V_{ca}(0)$, or $V_{ca}(t) \in \mathcal{L}_\infty$. Therefore, η, $\tilde{\mathbf{\Theta}}$, $\hat{\mathbf{\Theta}}$ and $E_b^- \in \mathcal{L}_\infty$, resulting in $w(x,t) \in \mathcal{L}_\infty([0,a])$ and $\int_0^a w_t^2 dx \in \mathcal{L}_\infty$. We assume that $w(x,t) \in \mathcal{S}_{\mathcal{A}_0}([0,a])$ so $w \in \mathcal{L}_\infty$ implies w_x, w_{xx}, w_{xxx} and $w_{xxxx} \in \mathcal{L}_\infty([0,a])$. We also assume that $\int_0^a w_t^2 dx \in \mathcal{L}_\infty$ implies w_t, w_{xt}, and $w_{xxt} \in \mathcal{L}_\infty([0,a])$. Thus, w_t, w_{xt}, w_{xxt}, w_x, w_{xx}, w_{xxx}, $w_{xxxx} \in \mathcal{L}_\infty([0,a])$. From Eq. (2.58), $w_{tt} \in \mathcal{L}_\infty([0,a])$ and we assume that this implies $w_{xtt} \in \mathcal{L}_\infty([0,a])$. From Lemma 55, $w_{xx}^+ \in \mathcal{L}_\infty$, so $\mathbf{W} \in \mathcal{L}_\infty$, and $\tau_c \in \mathcal{L}_\infty$, yielding $\dot{\eta} \in \mathcal{L}_\infty$. This results in

$$\dot{g}(t) = -\lambda_{3a}\left(\dot{E}_b + 2\eta\dot{\eta}\right) \in \mathcal{L}_\infty. \tag{6.67}$$

Thus, $g \to 0$ as $t \to \infty$, and $w(x,t) \to 0$, as $t \to \infty$, for all $x \in [0,a]$.
□

Experimental Verification

An axially moving belt control system implements the proposed control algorithms and compares them with standard PID control. Figure 18 shows the belt, drive motor, pulleys, pivoting roller, disturbance arm, and laser sensors of the experimental system. The beam model is appropriate for the moderately stiff belt operating under low tension. The drive motor and pulleys move the belt at 1.0 m/s during all tests. The excitation frequency ω equals the first natural frequency of the open loop system (7.2 Hz). A variety of excitation frequencies and amplitudes provide similar results. The disturbance arm provides repeatable, periodic forcing $f_u(t) = A\sin(\omega t)$ to the uncontrolled subdomain. The pivoting roller is actuated by a brushed DC motor that provides the control torque. The controllers (5.310) and (6.59) require measurement of the curvature on either side of the actuator, the curvature rate on the controlled side of the actuator, and the angular rate of the actuator. An encoder on the actuator motor measures the actuator angle. Backwards differencing of the encoder signal provides the actuator angular rate. Laser sensors measure the belt displacement on either side of the actuator. We assume the curvature is constant between the actuator and the laser sensor. Deviations of the belt displacement from a straight line between the actuator and sensor are then proportional to curvature. Backwards differencing of the controlled-side curvature provides w_{xxt}^{-}. A Mechatronics Workstation consisting of a Pentium 166 MHz Personal Computer (PC) running the QNX operating system and the software Qmotor implement the control algorithms. Filtered backwards differencing for velocity measurements and trapezoidal integration are used. Figure 19 shows the open loop response to an impulse disturbance near the controlled subdomain pulley provided by an impact hammer (d)-(f) and the periodic dancer arm disturbance input on the uncontrolled subdomain (a)-(c). The response of the controlled subdomain (a) has approximately 1.1 mm root mean squared (RMS) amplitude.

PID Control For comparison purposes, Fig. 20 shows the system response using simple PID control as follows:

$$\tau_c = K_p w_x(a,t) + K_d w_{xt}(a,t) + K_I \int_0^t w_x(a,\tau)\,d\tau, \tag{6.68}$$

where the control gains are designed to provide the best disturbance rejection to the periodic inputs. The PID control significantly reduces the controlled subdomain response (a) to approximately 0.70 mm RMS amplitude using the 0.45 V RMS control voltage (d) in response to the periodic disturbance. However, the PID controller reduces the damping associated with the impulse disturbance (c) using the control voltage (h) with a longer transient response time (1.0 s) than the open loop (0.7 s).

EMK Control Figure 21 shows the EMK control response to the same impulsive and periodic disturbances as the PID control experiment. The responses (a) for

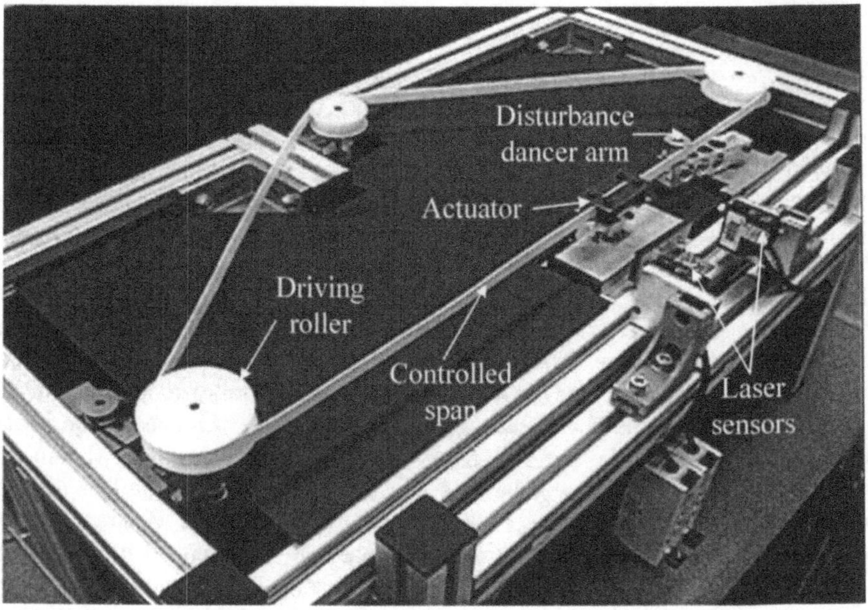

Figure 18 Axially moving beam experimental setup.

the controlled subdomain are less than 0.56 mm RMS amplitude to the periodic inputs using the similar control voltage (d) to PID control. The controlled subdomain transient resulting from impulse inputs (e) decays more than twice as fast (0.4 s), however, than PID, and almost twice as fast as open loop.

Adaptive Control Figure 22 shows the adaptive control response to the same disturbances as the PID control experiment. After an initial transient parameter convergence, the responses for controlled subdomain (a) are less than 0.62 mm RMS to the periodic inputs using similar control input (d) to PID control. Figure (e) and (f) show the convergence of the system parameters from zero to near the actual values of 0.0005 $N \cdot m^2$ and 0.0001 $kg \cdot m^2$ for EI and J, respectively. After parameter convergence, the impulse response is almost identical to the EMK result.

Results Discussion

The objectives of the proposed control approach are to isolate the controlled subdomain from bounded disturbances in the uncontrolled subdomain and damp vibration in the controlled subdomain. Heuristically, the controller acts as a one-way valve, confining vibration energy in the uncontrolled subdomain and eliminating vibration energy from the controlled subdomain. Pinning an axially moving beam using a

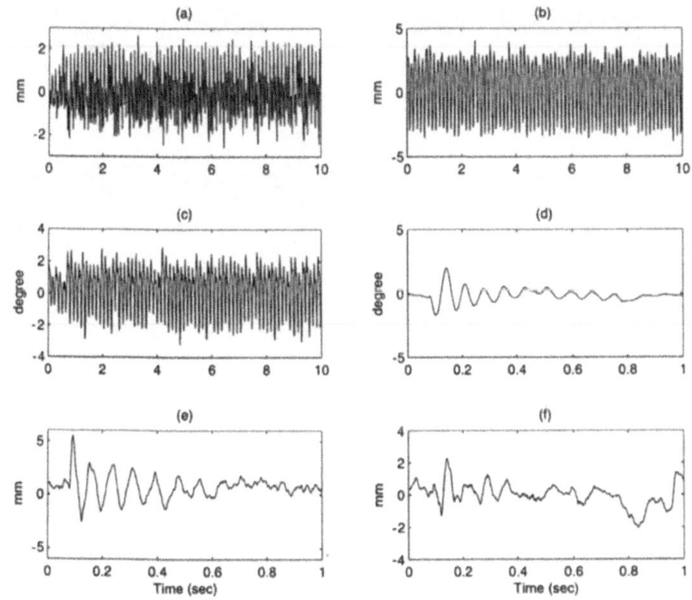

Figure 19 Open loop response: (a) Periodic disturbance response of $u(a^-, t)$ (controlled span); (b) Periodic disturbance response of $u(a^+, t)$ (uncontrolled span); (c) Periodic disturbance response of actuator angle $u_x(a, t)$; (d) Impulse response of actuator angle $u_x(a, t)$; (e) Impulse response of $u(a^-, t)$ (controlled span); (f) Impulse response of $u(a^+, t)$ (uncontrolled span).

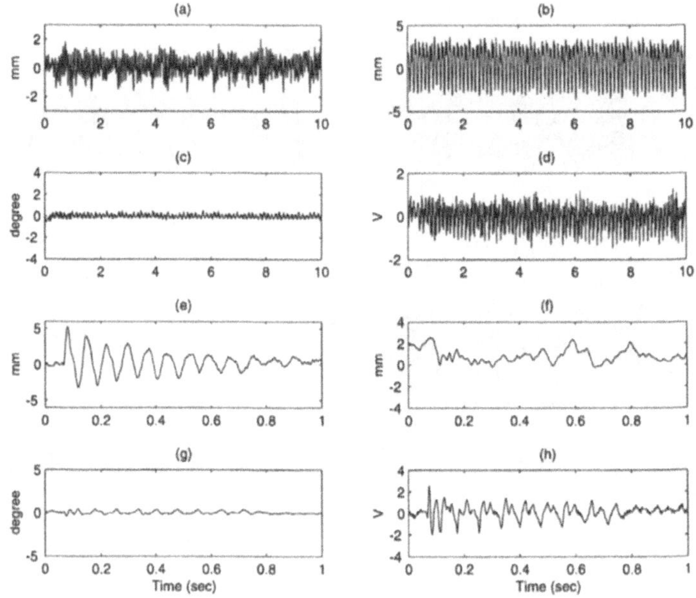

Figure 20 PID control response: (a) Periodic disturbance response of $u\left(a^{-},t\right)$ (controlled span); (b) Periodic disturbance response of $u\left(a^{+},t\right)$ (uncontrolled span); (c) Periodic disturbance response of actuator angle $u_x\left(a,t\right)$; (d) Control voltage for periodic disturbance; (e) Impulse response of $u\left(a^{-},t\right)$ (controlled span); (f) Impulse response of $u\left(a^{+},t\right)$ (uncontrolled span); (g) Impulse response of actuator angle $u_x\left(a,t\right)$; (h) Control voltage for impulse input.

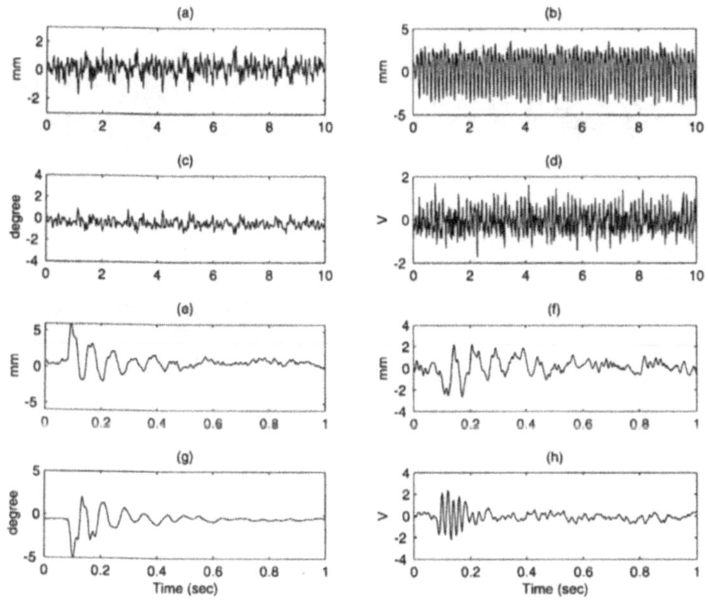

Figure 21 EMK control response: (a) Periodic disturbance response of $u(a^-,t)$ (controlled span); (b) Periodic disturbance response of $u(a^+,t)$ (uncontrolled span); (c) Periodic disturbance response of actuator angle $u_x(a,t)$; (d) Control voltage for periodic disturbance; (e) Impulse response of $u(a^-,t)$ (controlled span); (f) Impulse response of $u(a^+,t)$ (uncontrolled span); (g) Impulse response of actuator angle $u_x(a,t)$; (h) Control voltage for impulse input.

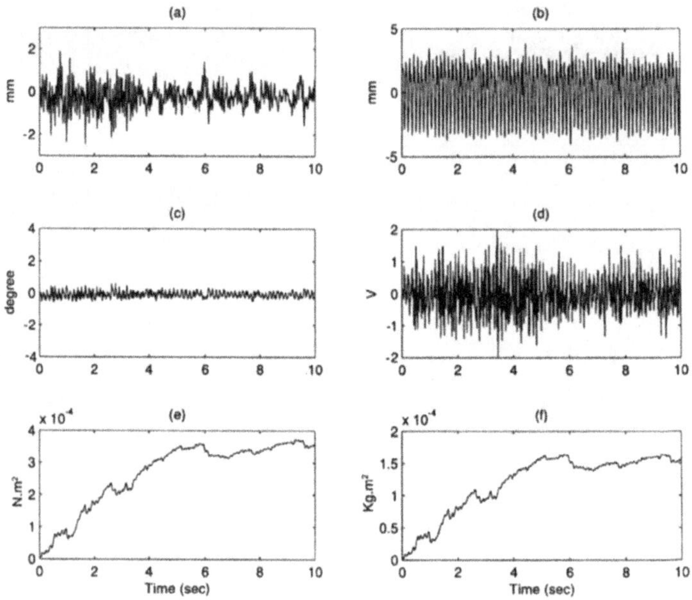

Figure 22 Adaptive control response for periodic disturbance: (a) Periodic disturbance response of $u(a^-,t)$ (controlled span); (b) Periodic disturbance response of $u(a^+,t)$ (uncontrolled span); (c) Periodic disturbance response of actuator angle $u_x(a,t)$; (d) Control voltage for periodic disturbance; (e) Estimator $\hat{E}I$; (f) Estimator $\hat{J}$

roller does not isolate adjacent subdomains unless the wrap angle around the roller is sufficiently large or multiple rollers are used to prevent vibration transmission via bending stiffness coupling. The actuator used in this research has multiple rollers that clamp the axially moving beam. If the actuator is locked in place, then the controlled subdomain is perfectly isolated from disturbances in the uncontrolled subdomain. Figure 20(c) shows that PID control effectively regulates the actuator angle to zero, eliminating the coupling between the two subdomains and isolating the controlled subdomain from disturbances in the uncontrolled subdomain. The vibration in the uncontrolled subdomain (Fig. 20(b)) is clearly larger than the controlled subdomain but some residual vibration remains. Vibration propagation around the backside of the belt loop and once-per-cycle excitation from the belt splice cause the residual vibration. The PID controller can be thought of as a best-case vibration isolator. It is therefore expected that the EMK and Adaptive controllers should have similar vibration isolation performance as PID control. The improved damping of the EMK and Adaptive controllers provides the slightly improved disturbance rejection over PID.

PID control reduces the damping in the controlled subdomain when compared to open loop (See Figs. 19(e) and 20(e)), violating a major control objective. In order to obtain best-case vibration isolation, the PID controller clamps the actuated end of the controlled subdomain, eliminating the energy dissipation associated with actuator friction. Figure 20(g) shows little actuator motion in response to the impulse input on the controlled subdomain. Reduction of the control gains increases vibration damping but reduces vibration isolation. The EMK and Adaptive controllers, however, provide much improved damping (Fig. 21(e)) without sacrificing isolation performance. The actuator moves in response to the impulse disturbance on the controlled side (Fig. 21(g)) but not the periodic disturbance on the uncontrolled side (Fig. 21(c)).

Chapter 7
BIBLIOGRAPHY

1. Abramowitz M, Stegun I (1965) *Handbook of Mathematical Functions.* Dover, New York
2. Baicu CF, Rahn CD, Nibali BD (1996) Active Boundary Control of Elastic Cables: Theory and Experiment. *J Sound and Vibration* 198:17-26
3. Baicu CF, Rahn CD, Dawson DM (1996) Boundary Control of a Flexible Link Electrically Driven Gantry Robot. *1996 ASME International Congress*, Atlanta, GA, DE-93:109-115
4. Balas MJ (1978) Modal control of certain flexible dynamic systems. *SIAM J Control and Optim* 16(3):450-462
5. Balakrishnan AV (1976) *Applied Functional Analysis.* Springer-Verlag, New York
6. Ball J, Slemrod M (1979) Nonharmonic Fourier Series and the Stabilization of Distributed Semi-Linear Control Systems. *Comm Pure Appl Math* 32:555-587
7. Berkman EF, Bender EK (1997) Perspectives on Active Noise and Vibration Control. *Sound and Vib* Jan:80-94
8. Birdsong CB, Radcliffe CJ (1999) A Compensated Acoustic Actuator for Systems with Strong Dynamic Pressure Coupling. *ASME Journal of Vibration and Acoustics* 121:89-94
9. Birkhoff G, Rota G-C (1962) *Ordinary Differential Equations*, Ginn And Company, New York
10. Biswas S, Ahmed N (1986) Stabilization of a Class of Hybrid Systems Arising in Flexible Spacecraft. *J Optim Theory and Applic* 50:83-108
11. Blair C (1995) *Integrator Backstepping Techniques for the Control of Loudspeaker Performance: Theory and Implementation*, Clemson University Thesis
12. Book WJ (1993) Controlled motion in an elastic world. *J Dyn Sys Meas Control* 115:252–26
13. Book WJ, Maizza-Neto O, Whitney DE (1975) Feedback control of two beam, two joint systems with distributed flexibility. *J Dyn Sys Meas Control* 97:424–431
14. Boustany F, d'Andréa-novel B (1992) Adaptive Control of an Overhead Crane using Dynamic Feedback Linearization and Estimation Design. *Proceedings of the 1992 IEEE International Conference on Robotics and Automation*, Nice, France, pp 1963-1968

15. Burke S, Hubbard J (1988) Distributed Actuator Control Design for Flexible Beams. *Automatica*, 24(5):619-627
16. Butler H, Honderd G, Van Amerongen J (1991) Model Reference Adaptive Control of a Gantry Crane Scale Model. *IEEE Control Systems*, pp 57-62
17. Canbolat H, Dawson D, Rahn C, Nagarkatti S (1998) Adaptive Boundary Control of Out-of-Plane Cable Vibration. *J Appl Mech* 65:963-969
18. Canbolat H, Dawson D, Rahn C, Vedagarbha P (1998) Boundary Control of a Cantilevered Flexible Beam with Point-Mass Dynamics at the Free End. *Mechatronics* 8(3):163-186
19. Carrier G (1945) On the non-linear vibration problem of the elastic string. *Quart Appl Math* 3:157-165
20. Chen C (1984) *Linear System Theory and Design.* Holt, Rinehart, and Winston New York
21. Chen G (1979) Energy decay estimates and exact boundary value controllability for the wave equation in a bounded domain. *J Mathématiques Pures et Appl* 9(58):249-274
22. Chen G, Delfour M, Krall A, Payre G (1987) Modeling, Stabilization, Control of Serially Connected Beams. *SIAM J Control Optim* 25(3):526–546
23. Chen G, Lee E, Littman W, Markus L (eds) (1991) Distributed Parameter Control Systems New Trends and Applications. Marcel Dekker New York
24. Choura S (1995) Control of Flexible Structures With the Confinement of Vibrations. *J Dyn Sys Meas Control* 117:155-164
25. Clark RL, Cole DG (1995) Active Damping of Enclosure Sound Fields Through Direct Rate Feedback Control. *J Acoust Soc Am* 97(3):1710-1716
26. Clark RL, Frampton KD (1996) Phase Compensation for Feedback Control of Enclosed Sound Fields. *J Sound and Vib* 195(5):701-708
27. Clark WW, Robertshaw HH (1997) Force Feedback in Adaptive Trusses for Vibration Isolation in Flexible Structures. *J Dyn Sys Meas Control* 119:365-371
28. Curtain RF, Zwart HJ (1995) *An Introduction to Infinite-Dimensional Linear Systems Theory.* Springer-Verlag, New York
29. Dafermos CM, Slemrod M (1973) Asymptotic Behavior of Nonlinear Contraction Semigroups. *J Funct Anal* 13:97-106
30. d'Andréa-novel B, Boustany F, Conrad F, Rao B (1994) Feedback Stabilization of a Hybrid PDE-ODE System: Application to an Overhead Crane. *Math Control Sig Sys* 7:1-22
31. de Queiroz M, Dawson D (2000) *Liapunov-Based Control of Mechanical Systems.* Birkhauser, Boston
32. Doak PE, (1973) Excitation, Transmission and Radiation of Sound From Source Distributions in Hard-Walled Ducts of Finite Length (I): The Effects of Duct Cross-Section Geometry and Source Distributions Space-Time Pattern. *J Sound and Vib* 31(1):1-72

33. Ebeid AM, Moustafa KAF, Emara-Shabaik HE (1992) Electromechanical Modeling of Overhead Cranes. *Int J Sys Sci* 23(12):2155-2169
34. Electric Power Research Institute (1979) *Transmission Line Reference Book, Wind-Induced Conductor Motion.* Electric Power Research Institute, Palo Alto, CA
35. Elliot SJ, Stothers IM, Nelson PA (1987) A multiple error LMS algorithm and its applications to the active control of sound and vibration. *IEEE Trans Acoust Speech Sig Proc* ASSP-35:1423-1434
36. Ertur D, Li Y, Rahn, CD (1998) Adaptive Vibration Isolation for Flexible Structures. *1998 ASME International Congress & Exhibition*
37. Feng D, Shi D, Zhang W (1996) Feedback Stabilization of Timoshenko Beam with Boundary Dissipation. *Proc 1996 IFAC 13th Triennial World Congress*, pp 311-316
38. Fliess M, Levine J, Rouchon P (1991) A Simplified Approach of Crane Control Via a Generalized State-Space Model. *Proceedings of the 30th Conference on Decision And Control*, Brighton, England, pp 736-741
39. Franklin GF, Powell JD, Workman M (1998) *Digital Control of Dynamic Systems*, Addison Wesley Longman, Menlo Park, CA
40. Fried I (1982) Large Deformation Static and Dynamic Finite Element Analysis of Extensible Cables. *Comp and Struct* 15:315-319
41. Fujino Y, Warnitchai P, Pacheco B (1993) Active Stiffness Control of Cable Vibration. *J Appl Mech* 60:948-953
42. Fujino Y, Susumpow T (1995) Active Control of Multimodal Cable Vibrations by Axial Support Motion. *Smart Matls and Struct* 4:41-51
43. Fuller CR, Elliot SJ, Nelson PA (1996) *Active Control of Vibration.* Academic Press, London
44. Fung R-F, Wu J-W, Wu S-L (1999) Exponential stabilization of an axially moving string by linear boundary feedback. *Automatica* 35:177 - 181
45. Gawronski W (1996) *Balanced Control of Flexible Structures.* Springer-Verlag, London
46. Gawronski, W (1998) *Dynamics and Control of Structures a Modal Approach.* Springer-Verlag, New York
47. Gould LA, Murray-Lasso MA (1966) On the Modal Control of Distributed Systems with Distributed Feedback. *IEEE Trans Auto Control* AC-11(4):729-737
48. Guigou C, Fuller CR, Wagstaff PR (1994) Active Isolation of Vibration with Adaptive Structures. *J Acoust Soc Amer* 96(1):294-299
49. Hahn W (1967) *Stability of Motion.* Springer-Verlag, New York
50. Hardy G, Littlewood J, Polya G (1959) *Inequalities.* Cambridge University Press, Cambridge
51. Hillerström G (1996) Adaptive Suppression of Vibrations–A Repetitive Control Approach. *IEEE Trans Control Sys Tech* 4(1):72-78
52. Hull AJ, Radcliffe CJ, Southward SC (1993) Global Active Noise Control of a One-Dimensional Acoustic Duct Using a Feedback Controller. *J Dyn Sys Meas Control*

115:488-494

53. Ince EL (1926) *Ordinary Differential Equations.* Dover, New York
54. Inman DJ (1989) *Vibration with Control, Measurement, Stability.* Prentice Hall, Englewood Cliffs, NJ
55. Irvine H (1981) *Cable Structures.* MIT Press, Cambridge, MA
56. Jenkins MD, Nelson PA, Pinnington RJ, Elliott SJ (1993) Active Isolation of Periodic Machinery Vibrations. *J Sound and Vib* 166(1):117-140
57. Jones JF, Petterson BJ (1988) Oscillation Damped Movement of Suspended Objects. *Proceedings of IEEE International Conference on Robotics and Automation*
58. Joshi S, Rahn CD (1995) Position control of a flexible cable gantry crane: theory and experiment. *Proceedings of the American Control Conference*, Seattle, WA, pp 2820-2824
59. Junkins JL, Kim Y (1993) *Introduction to Dynamics and Control of Flexible Structures*, AIAA, Washington, DC
60. Karnopp D (1995) Active and Semi-Active Vibration Isolation. *Trans ASME* 117:177-185
61. Kato T (1980) *Perturbation Theory of Linear Operators.* Springer-Verlag, New York
62. Kinsler LE, Frey AR, Coppens AB, Sanders JV (1982) Fundamentals of Acoustics. John Wiley & Sons New York
63. Kokotovic P (1992) The Joy of Feedback: Nonlinear and Adaptive. *IEEE Control Sys Mag* 12:7-17
64. Kolmogorov AN, Fomin SV (1975) *Introductory Real Analysis.* Dover Publications, New York
65. Komornik V (1991) Rapid boundary stabilization of the wave equation. *SIAM J Control and Optim* 29(1):197-208
66. Krstić M, Kanellakopoulos I, Kokotović P (1995) *Nonlinear and Adaptive Control Design*, John Wiley and Sons, New York
67. Lagnese J (1988) Note on boundary stabilization of wave equations. *SIAM J Control Optim* 26(5):1250-1256
68. Lane SA, Clark RL (1998) Dissipative Feedback Control of a Reverberant Enclosure Using a Constant Volume Velocity Source. *J Vib Acoust* 120:987-993
69. Lasiecka I (1989) Stabilization of wave and plate-like equations with nonlinear dissipation on the boundary. *J Diff Eqs* 79:340-381
70. Lee C-K, Moon FC (1990) Modal Sensors/Actuators. *J Appl Mech* 57:434-441
71. Lee SY, Mote CD, Jr. (1996) Vibration control of an axially moving string by boundary control. *J Dyn Sys Meas Control* 118:66-74
72. Leipholz HHE (1980) *Stability of Elastic Systems.* Sijthoff and Noordhoff, Germantown MD
73. Liu B, Tzou H-S (1998) Distributed Photostrictive Actuation and OptoPiezothermoelasticity Applied to Vibration Control of Plates. *J Vib Acoust* 120:937-943

74. Luo ZH, Guo BZ (1997) Shear Force Feedback Control of a Single-Link Flexible Robot with a Revolute Joint. *IEEE Trans Auto Control* 42(1):53–65
75. Luo Z (1993) Direct Strain Feedback Control of Flexible Robot Arms: New Theoretical and Experimental Results. *IEEE Trans Auto Control* 38(11):1610–1622
76. Luo Z, Guo B (1995) Further Theoretical Results on Direct Strain Feedback Control of Flexible Robot Arms. *IEEE Trans Auto Control* 40(4):747–751
77. Luo Z, Kitamura N, Guo B (1995) Shear Force Feedback Control of Flexible Robot Arms. *IEEE Trans Rob and Auto* 11(5):pp 760–765
78. Luo Z, Guo B, Morgül O (1999) *Stability and Stabilization of Infinite Dimensional Systems with Applications.* Springer-Verlag, London
79. Liapunov MA (1949) *Problème Général de la Stabilité du Mouvement*, Princeton Univ Press, Princeton
80. Martindale S, Dawson D, Zhu J, Rahn C (1995) Approximate Nonlinear Control for a Two Degree of Freedom Overhead Crane: Theory and Experimentation. *Proc of the American Control Conference*, Seattle, WA, pp 301-305
81. Meirovitch L (1967) *Analytical Methods in Vibrations.* MacMillan, New York
82. Meirovitch L (1980) *Computational Methods in Structural Dynamics.* Sijthoff & Noordhoff, Alphen aan den Rijn, Netherlands
83. Meirovitch L (1990) *Dynamics and Control of Structures*, John Wiley & Sons, New York
84. Morgül O (1994) A dynamic control law for the wave equation. *Automatica* 30:1785–1792
85. Morgül O (1991) Orientation and Stabilization of a Flexible Beam Attached to a Rigid Body: Planar Motion. *IEEE Trans Auto Control* 36(8):953-962
86. Morgül O (1992) Dynamic Boundary Control of a Euler-Bernoulli Beam. *IEEE Trans Auto Control* 37(5):639-642
87. Morgül O (1992) Control and Stabilization of a Rotating Flexible Structure. *Automatica* 30(2):351-356
88. Morgül O (1990) Control and Stabilization of a Flexible Beam Attached to a Rigid Body. *Int J Control* 51(1):11-31
89. Morgül Ö, Rao B, Conrad F (1994) On the Stabilization of a Cable with a Tip Mass. *IEEE Trans Auto Control* 39(10):2140-2145
90. Moustafa K, Ebeid A (1988) Nonlinear Modeling and Control of Overhead Crane Load Sway. *J Dyn Sys Meas Control* 110:266-271
91. Movchan A (1959) The Direct Method of Liapunov in Stability Problems of Elastic Systems. *J App Math Mech* 23(1): 686-700
92. Nagarkatti S, Dawson D, de Queiroz M, Costic B (1998) Boundary Control of a Two-Dimensional Flexible Rotor. *Proc IEEE Conf Decision and Control* pp 2581-2586, Tampa, FL
93. Nagarkatti S, Zhang F, Rahn C, Dawson D (1999) Backstepping Boundary Control for Noise Reduction in Finite Ducts. *IEEE/ASME International Conference on*

Advanced Intelligent Mechatronics, Atlanta, GA, pp 836-841

94. Nelson PA, Elliot SJ (1992) *Active Control of Sound*, Academic Press, New York
95. O'Connor L (1994) Generating the sounds of silence. *Mechanical Engineering* April:54-58
96. Oplinger D (1960) Frequency response of a nonlinear stretched string. *J Acoust Soc Am* 32:1529-1538
97. Olson H, May E (1953) Electronic Sound Absorber. *J Acoust Soc Amer* 25:1130-1136
98. Porter B, Crossley R (1972) *Modal Control.* Taylor and Francis, London
99. Pazy A (1983) *Semigroups of Linear Operators and App lications to Partial Differential Equations.* Springer-Verlag, New York
100. Pedersen M (1999) *Functional Analysis in Applied Mathematics and Engineering.* Chapman & Hall, New York
101. Perkins N, Mote C (1987) Three-dimensional Vibration of Travelling Elastic Cables. *J Sound Vib* 114(2):325-340
102. Piccinini L, Stampacchia G, Vidossich G (1984) *Ordinary Differential Equations in* R^n, Springer-Verlag, New York
103. Preumont A (1997) *Vibration Control of Active Structures*, Kluwer Academic, Dordrecht
104. Costes, N (2000) Qmotor 3.0, Quality Real-Time Systems, Falls Church, VA
105. Queiroz M, Dawson D, Rahn C, Zhang F (1997) Vibration Control of a Distributed Axially Moving String Using Two Actuators. *1997 ASME Design Engineering Technical Conferences*, Sacramento, DETC97/VIB-3781
106. Quinn J, Russell D (1977) Asymptotic stability and energy decay rates for solutions of hyperbolic equations with boundary damping. *Proceedings of Royal Society* **A 77**: 97–127
107. Renshaw A, Rahn C, Wickert J, Mote C (1998) Energy and Conserved Functionals for Axially Moving Materials. *J Vib Acoust* 120(2):634-636
108. Radcliffe C, Gogate S, Hall A (1994) Development of Active Acoustic Sink (AAS) for Noise Control Applications. *Active Control of Vibrations and Noise*, ASME
109. Radcliffe C, Gogate S (1992) Identification and Modeling Speaker Dynamics for Acoustic Control Applications. *ASME Symposium on Active Control of Noise and Vibration*
110. Rahn C, Mote, C (1993) Axial Force Stabilization of Transverse Beam Vibration. *Proceedings of the* 14th *ASME Biennial Conference on Mechanical Vibration and Noise,* Albuquerque, NM, DE-61:29–34
111. Seta W (1971) *Theory and Problems of Acoustics*, McGraw-Hill, New York
112. Sakawa Y, Nakazumi A (1985) Modeling and Control of a Rotary Crane. *J Dyn Sys Meas Control* 107:200-206
113. Sakawa Y, Shindo Y, Hashimoto Y (1981) Optimal Control of a Rotary Crane. *J Optim Theory Applic* 35(4):535-557

114. Sallet G, Xu C, Laousy H (1995) Boundary Feedback Stabilization of a Rotating Body-Beam System. *Proc of the IEEE Conf on Decision and Control*, New Orleans, LA, pp 930-935
115. Saperstone S (1981) *Semidynamical Systems in Infinite Dimensional Spaces*, Springer-Verlag, New York
116. Shahruz M, Krishna L (1996) Boundary control of a nonlinear string. *J Sound and Vib* 195:169-174
117. Shahruz M (1997) Suppression of Vibration in a Nonlinear Axially Moving String by Boundary Control. *ASME Design Engineering Technical Conferences* 106(1):6-14
118. Slotine J, Li W (1991) *App lied no nlinear Control.* Prentice Hall, Englewood Cliffs, NJ
119. Soler A (1970) Dynamic Response of Single Cable with Initial Sag. *J Frank Inst* 190:377-387
120. Stakgold I (1979) *Green's Functions and Boundary Value Problems*, John Wiley & Sons, New York
121. Starr G (1985) Swing-Free Transport of Suspended Objects With a Path-Controlled Robot Manipulator. *J Dyn Sys Meas Control* 107:97-100
122. Steinman D, Watson S (1957) *Bridges and Their Builders.* Dover, New York
123. Tanabe H (1979) *Equations of Evolution*, Pitman, New York
124. Taylor A (1985) *General Theory of Functions and Integration.* Dover, New York
125. Tzou H, Bergman L (eds) (1998) *Dynamics and Control of Distributed Systems*, Cambridge University Press, London
126. Tzou H, Tseng C (1990) Distributed Piezoelectric Sensor/Actuator Design for Dynamic Measurement/Control of Distributed Parameter Systems: A Piezoelectric Finite Element Approach. *J Sound Vib* 138(1):17-34
127. Ulsoy A (1984) Vibration Control in Rotating or Translating Elastic Systems. *J. Dyn Sys Meas Control* 106(1):6-14
128. West H, Geschwinder L, Suchoski J (1975) Natural Vibrations of Suspension Cables. *American Society of Civil Engineers, Journal of the Structural Division* 101:2277-2291
129. Xu C, Baillieul, J (1993) Stabilizability and stabilization of rotating body-beam system with torque control. *IEEE Trans Auto Control* 38:1754-1765
130. Yang B, Mote C (1991) Active Vibration Control of the Axially Moving String in the S Domain. *J Appl Mech* 58:189-196
131. Yosida K (1978) *Functional Analysis*, Springer-Verlag, London
132. Zuazua E (1990) Uniform stability of the wave equation by nonlinear boundary feedback. *SIAM J Control Optimn* 28(2):466-477
133. Zhang F, Dawson D, Rahn C (1997) Boundary Control of a Cantilevered Timoshenko Beam with Mass/Inertial Dynamics at the Free End. *IEEE Conf on Decision and Control*, San Diego, CA

Index

Printing: Saladruck, Berlin
Binding: H. Stürtz AG, Würzburg

Batch number: 10682128

Printed by Printforce, the Netherlands